WE OWN 29%

ET

HAS THE REST

Martin Thomas

ISBN: 978-1-83709-256-7

Previous publication by the author:

UFO – Friend or Foe? August 2025

This work is dedicated to the BLAZE Television Channel for making me think, to Google for helping me find information and to Wikipedia for denying everything.

The Scientific Method

Develop an hypothesis which can be used to make testable predictions. Test these predictions and if they work, test again for repeatability.

The Engineering Method

If it walks like a duck, and quacks like a duck, it probably is a duck.

WE OWN 29% - ET HAS THE REST

Table of Contents

LIST OF FIGURES & TABLES

PREFACE

In my first book "UFO – Friend or Foe?[1]", I made a broad-ranging sweep through the history of sightings across as much of our planet as I could.

In fact the terms Unidentified Flying Object (UFO) and Unidentified Submersible Object (USO) have been merged as Unidentified Anomalous Phenomena (UAPs). I will be using the terms UAP and submersible UAPs throughout this book.

The term "Alien" has been used to describe any occupants or manufacturers of UAPs. However, this word has many pejorative associations – it means basically "not one of us" and is frequently used to describe immigrants, whose presence may be illegal, or who are disliked because of their race, colour or creed. I intend to use the word "Others" throughout this book - perhaps not an ideal term, but at least it avoids many of the pre-assumptions inherent in traditional terms.

In reviewing evidence of all the sightings of UAPs, I was easily able to convince myself, and hopefully my readers, that they really existed, and that they had some bases underground, and underwater. There could easily be 20 or 30 such bases, judging from the distribution of UAP sightings. It is likely that this is an under-estimate but, even so, this does seem excessive if all they are doing is stopping us from destroying our planet with atom bombs, gathering rare earths, and studying our genetics.

What we are seeing is colonisation, with "Other" species sharing the planet with us, with only their over-flights intruding on our consciousness. The answer to malevolent "Others" is to have as many fully autonomous colonies as possible, and to keep them small. This is what is happening on Earth, although it is not clear how many are involved.

So there is every chance that many of the underwater bases could belong to Extra-Terrestrials (ETs). They may be water-breathers, but there is no reason why they should not be species requiring a gaseous atmosphere, living in an artificial biosphere. Thus they could also be other earth-evolved species – Intra-Terrestrials (ITs), members of an earlier human race, or even entities from another parallel dimension (PDEs).

I concluded my first book with the following statements:

> Yes, there are UAPs in our skies, and yes "Others" exist. Most of these UAPs are either small un-crewed reconnaissance drones or are crewed by ETs. They appear to be fighting a desperate battle to save our planet, even at risk of their own lives.
>
> Of course, it is not just humans that they are trying to save. There are also an unknown number of "Other" colonies in our mountains and under the seas.

Chapter 1

INTRODUCTION

a) EARLY TIMES

The last ice-age started about 2.6 million years BCE and finished about 11,700 years BCE. It must be remembered, though, that a large proportion of the land-mass of our planet was not covered by ice-sheets, although it was a lot colder. There are indications of civilisation going back in Gobekli Tepe[2] in Turkey to at least 12,000 BCE, but there are certain features, such as carvings of figures with six fingers, which suggest they may not have belonged to our current human civilisation.

The first signs of our present human civilisation are thought to occur at roughly 6,500 BCE in Sumeria. There are legends of gods called Anannaki[3] at that time, teaching medicine, agriculture, animal husbandry, astronomy and mathematics. The wheel was first adopted about then.

These gods are the first recorded contact with "Others" by our present civilisation. Of course this does not mean that this was the first arrival of these "Others" on our planet. Indeed, there must have been at least one prior arrival if the hypothetical genetic anomaly[4] in our evolution is to be believed.

However, as I aimed to demonstrate in my previous book, "Others" have been colonists here, perhaps for millennia, living on mountain tops, in caves, or in artificial bases underwater in lakes or the sea.

The reason I have already mentioned the ice-age is that the vast northern ice-sheets could have determined, to some extent, where these colonising "Others" may have set up camp. (See Figure 1)

Even within the ice-sheets, there would still have been massive mountains rising above the ice, offering potential base sites.
In the lands away from the ice sheets, they could have come at any time. There would have been plenty of choice of mountains, lakes or oceans, even thought the sea level would have been perhaps 120 metres lower[5] then.

b) THE GODS ARRIVE

I am going to be concentrating in this book on underwater bases, although there are certainly still some mountain bases such as Mount Hayes[6] in Alaska, Mount Shasta[7] in California and Mount Kailash[8] in Tibet as well as volcanoes such as Mount Shishaldin[9] in the Aleutian Islands and Mount Popacatepétl[10] in Mexico,

If some early colonists assisted in the development of various human tribes, they could have come from the oceans.

There are many oral histories of "Others", taken then to be gods, visiting tribes to teach such basic things as agriculture, cooking, fishing, medicine and building.

FIGURE 1. Map of the world showing sea levels and ice sheets at the peak of the last ice age.

Some of these are specifically described as sea gods, even though they may have arrived in flaming chariots:

- **Oannes**,[11] in Mesopotamian mythology, was an amphibious being who taught mankind wisdom. He had the form of a fish but with the head of a man under his fish's head and under his fish's tail the feet of a man. In the daytime he came up to the seashore of the Persian Gulf and instructed mankind in writing, the arts, and the sciences.
- **The Nommo**[12] are ancestral spirits (sometimes referred to as deities) worshipped by the Dogon tribe of Mali, Africa. The word Nommo is derived from a Dogon word meaning: "to make one drink". Folk art depictions of the Nommos show creatures with humanoid upper torsos, legs/feet, and a fish-like lower torso and tail. The Nommos are also referred to as "The Masters of the Water", "The Monitors", and "The Teachers". The Dogon legends state that the Nommos required a watery environment in which to live.
- **Dagon**[13] was the god of the Philistines. The idol was represented in the combination of both man and fish. The name 'Dagon' is derived from 'dag' which means 'fish'. His name is a lot like 'Dogon'. Dagon was worshipped by the early Amorites, by the people of Ebla, by the people of Ugarit and a chief god (perhaps the chief god) of the Biblical Philistines.
- **Greek Sea Gods** It would appear that, as well as Titans, the Greek gods included many Sea Gods such as Amphitrite, Poseidon, and Triton as well as countless Sirens and Nymphs.

I think it is significant that the claimed Russian KGB list of "Others", which has supposedly been translated into English, makes no mention of amphibious aliens, even though there are records of possible sightings in Russia in other reports.

In 1982, in Lake Baikal[14], Russian divers came across underwater "Others" and attempted to capture them. This resulted in the deaths of three of the five divers. The "Others" were described as basically humanoid but adapted for the water.

In another book[15] listing "Other" races, there are references to several kinds of amphibians. There is one which is very similar to humans, but with webbed hands and feet. A second is more reptilian, and is believed to have returned to the sea after being land-dwelling. This book also refers to the Alabrans, who need to hydrate themselves regularly.

c) OR WERE THEY THERE ALREADY?

Although the great flood, which is recorded in the Bible and other east Mediterranean texts, is generally assumed to be the Mediterranean Sea breaking into the Black Sea, there is folklore of a Great Flood, sufficient to destroy all human society, in many human cultures around the world, and they would never have even heard of the Black Sea.

There are submerged coastal cities dated from the end of the ice-age, which are considered to be proof of this. As all the ice melted, it raised the sea level, leaving these cities under-water. However, the amount of water, necessary to destroy all human society would have far exceeded the melt-water from the end of the ice-age, so some have suggested that perhaps it came from some extra-terrestrial source such as water-filled planetoids. In this case, where did it all go to afterwards?

These submerged cities have been attributed to a part of the human race which developed earlier than us, and which is now called the Telosians. Alternatively, construction of these cities is laid at the feet of "Others", both ETs and ITs, and to the present human race. It reality, it is probably a mixture of all four.

As these cities, which were originally on dry land, were being destroyed, it resulted in considerable disruption to the stumbling development of humanity. This was probably the opportunity for "Others" to move in unobserved.

The end result is that, underneath our seas today, there are a number of colonies, hidden from human sight, where various "Others" can now live out their lives. These are:-

- Benevolent "Other" military bases, focused on protecting Earth from malevolent "Others" and, later on, preventing humanity from miss-using its atomic capabilities.
- Small "Other" colonies, focused on forming micro-versions of their home planetary societies to preserve their species in case of catastrophe.
- Intra-Terrestrials, living in isolation from their land-based neighbours.
- Perhaps residual populations of that earlier human race, the Telosians, who are claimed to be the descendents of Lemuria, a sunken land similar to Atlantis. However, some claim that they were wiped out, millennia ago.

Chapter 2

WHERE DID THE INTRA-TERRESTRIALS COME FROM?

a) GENERAL

The aquatic dinosaurs supposedly did not survive the land dinosaur extinction event. It has been suggested that the seas became too acidic for them. However, it is evident that other aquatic creatures did survive – the sea dinosaurs had to eat something, and not all these "somethings" can have disappeared.

b) FISH

To survive predation by the aquatic dinosaurs, one can speculate that it would have been an advantage for fish to have the ability to plan and to communicate with others of the same species. Some species could have developed more intelligence that we give them credit for. Survival has always been a driver for development.

The top of this food chain are the sharks. They are fish whose skeletons are made from cartilage rather than bone. They are part of a group including skates, rays, sawfish and over 350 species of shark.

We know of species of these which demonstrate remarkable co-operative hunting abilities and the most intelligent might have learnt how to avoid humans entirely as a survival technique. We may never know when or if they developed abilities beyond our own.

c) CETACEANS

In his book "The Naked Ape[16]", Desmond Morris suggested that, at one stage, early in their development, humans dwelt on shorelines, foraging for fish, shellfish, seaweed etc. To offer proof, he points out that what hair we and some water mammals have on our bodies is very different from other animals. It is streamlined from head to toe, perhaps to offer less resistance when swimming.

If humans were one-time beachcombers, it is not beyond the realms of possibility that two strains diverged here, leaving no fossil record. Some strains opted to remain in the water, remaining oxygen-breathing mammals, and evolving into dolphins, porpoises and whales.

We know that some species of cetaceans are very intelligent, and one can speculate that one deep-diving group might have become far brighter than us. Some whales can dive extremely deep, and stay under for long periods. They may have managed to avoid us entirely.

In the 1950s, John Lilly's[17] research helped to advance the view of dolphins as intelligent and socially aware creatures, a perspective that was a marked contrast to earlier, negative views of them. They have been shown to be quick learners and innovative, capable of teaching skills to others.

Even if they do not possess hands to let them become technologically developed, who can say how far their mental abilities could have advanced?

d) CEPHALOPODS

A cephalopod is any member of a class including squid, octopus, cuttlefish, or nautilus. They are widely regarded as the most intelligent of the invertebrates and have well-developed senses and large brains. The nervous system of cephalopods is the most complex of the invertebrates.

The sequencing of a full cephalopod genome has remained challenging to researchers due to the length and repetition of their DNA[18]. The genome showed similar patterns to other marine invertebrates with significant additions assumed to be unique to cephalopods. Cephalopod genomes are generally large, with some, like the squid, being larger than the human genome. Studies have revealed large-scale genomic rearrangements, which likely contribute to cephalopod morphological innovations.

Research has shown that octopuses can form concepts, plan for the future, generate a cognitive map, self-monitor for apparent pain and manipulate communication.[19]

Scientists have had great difficulty in fitting these creatures into any family tree, and it has been suggested that this is because they did not originate on Earth, although they may have been here for millions of years.

Nevertheless, they are here now, are extremely intelligent and have the ability to manipulate their environment. Not needing to breathe air, they could dwell deep underwater and could be an IT species, which we have never detected.

e) <u>ET ALIA</u>

If humans managed to evolve from small underground creatures in the time since the extinction of the dinosaurs, then it is not beyond the realms of possibility that other mammals, or even birds, which are descendants of dinosaurs, did as well, particularly if they had a running start on us.

Equally, there are creatures like reptiles, such as crocodiles, alligators, lizards and iguana still alive which, although not descended from dinosaurs, are historically related to them. It might be that some of their relatives could have out-evolved us.

The only thing missing is an appropriate fossil record. Perhaps we will find one one-day. To some extent, this will depend on whether individual species were capable of developing manipulatory appendages. If not, perhaps the species is in plain sight today, with unrecognised mental capacity. For now, we are specifically interested in sea dwelling creatures, but who knows?

In reality, before they can occupy our oceans undetected, ITs need either to be able to breathe underwater, or to have the ability to construct underwater habitats. That is not to say that ITs cannot occupy hidden places on the land. Native Americans talk of underground Ant People[20], and these are represented in petroglyphs in many places around the world.

Perhaps the best bet for an underwater IT is a cephalopod. The ones we know are very intelligent and equipped with more manipulatory appendages that we have. Who can say whether there is a more technically advanced version of these living in the deeps?

Beyond that, any amphibian or non-hoofed mammal could evolve into an advanced technological species, capable of building its own underwater habitat.

Chapter 3

WHERE ARE THE MOST LIKELY BASES?

a) GENERAL

One place to start is to look at the frequency of submersible UAP sightings and coastal sightings of "Others". In the first volume of his magnum opus, Richard Dolan[21] looks at all the credible sightings of USOs up to 1969 – a total of 178. His future two volumes will cover the remaining 494 up to the year 2024. For the moment, I will have to confine my analysis to the data in volume one and, where possible, combine this with data from other sources.

I have chosen to exclude certain classes of sightings from my analysis. First and foremost, I have virtually always excluded sightings where there is only one witness. Also, it has been claimed that UAPs hide inside clouds and volcanic smoke, but these dark spots could easily be denser patches and, too often, what are clearly lenticular clouds are claimed to be UAP mother-ships. Where sightings consist of spotting a single stationary night-time light, these are the most frequently dismissed as miss-identifications.

Richard Dolan's analysis of each sighting includes date, location and credibility, as well as what other facts he can glean from the sighting report. His credibility rating is particularly valuable, and is sadly lacking from many other sources. He breaks the credibility rating down into:

Weak	Can think of easier explanations
Moderate	
Solid	
Strong	
Certain	Absolutely no doubt

He does council caution in looking at UAP sightings, however. He describes in another book[22] how in Illinois in 2008, he spent 2 hours looking at the sky using high quality night-vision binoculars. In that time, he reckoned he saw about 100 high-flying UAPs. He doesn't attribute these just to "Others". He suspects that many were advanced USA craft, stemming from their reverse-engineering project.

Even so, I will start my analysis by looking at the reported sightings near to recognized water-based hot spots, and seek to identify any others that stand out. This analysis will become more accurate when Richard Dolan publishes his later volumes.

It must be born in mind that the last ice-age ended about 10,000 years ago, and the folklore of many races speaks of being visited by "Others" before then. At that time much of North America, Northern Europe and Russia was under vast ice-sheets, with only mountains protruding (See Figure 1). Whilst it is not impossible that “Others” established colonies in the lands under these ice-sheets, it would have been easier for them to start with those areas which were not frozen, and with mountain tops.

At the same time, I can look for any evidence of the type of “Other” species occupying the colonies.

I have broken down my analysis by maritime zones. The data is collated in tables in the Appendix.

In the Table on the next page, I have summarised the types of “Others” encountered.

DESCRIPTION	NAME	ORIGIN/ NAME	LOCATIONS	ACTIVITIES
2.0mTall & Hairy	Giant Apes	Yeti	Worldwide	Solitary existence in forests
1.3m-1.5m Big Head	Small Grey	Reticulum	Sweden & USA	Abductions & Medical Tests
1.6m Grey	Tall Grey	Reticulum	Sweden & USA	Assists Small Greys
2.5m Thin White	Very Tall White	Arcturian	South America & Worldwide	General Peacemaker
2.2m Fair Hair	Nordic	Apunians	South America & Worldwide	Helping humans
2.2m Fair Hair Webbed hands	Nordic	Cassiopeians	Northern Asia	Abductions
1.2m - 1.3m	Dwarfs		Wide spread	Many sorts Very ancient
0.1m – 1.0m	Little People		Wide spread	Many sorts Very ancient
2m Black Bird	Mothman		Widespread	Solitary
2m Bird	Avian		North Japan	Ancient
Lizards			Underground	Abductions
Tall Red/Orange			South America	Unknown
No necks			South America	Unknown
Cyclops			Americas	
Cats	Felines	Urmahs	Australia	
Dogs	Canines		Australia	Thylacine
Chupacabra	Goatsucker		South America	"Others" or "Others pets?
Human 1.8m	Earlier Humans	Telosians	Mt Shastna, Antartica	Unknown

Table 1 A summary of the land-based "Others" sightings reported in this book.

b) INDIAN OCEAN

i. Persian Gulf

This is the nearest potential undersea site to Sumaria the cradle of early human civilisation, where the Anannaki are reported to have nurtured our early civilisation, along with the Avians[23]. It is claimed that the Anunnaki were responsible for the genetic anomaly[24] in human development, 40,000 years ago, by introducing what is called the D allele to human DNA[25].

It is interesting that all the sightings up to 1956 are of this so-called pinwheel. This has not been fully explained but is reported as an under-water light effect, where beams of light seem to radiate from a central point and all stop at much the same distance from this. It is thought that it may be a bio-luminescent effect caused by sound waves, though no-one knows the source of the sound waves. A similar effect is reported in the Malacca Straight in Malaya and the Gulf of Carpentaria in Australia. These are all tropical seas.

I have been unable to find records of any sightings of any sort in the Persian Gulf subsequent to this, so I await Richard Dolan's next volume with interest.

It is claimed that the Anannaki left Earth about 5,000 years ago, presumably leaving their base. This would have either been in the Persian Gulf or in Iraq or Afghanistan.

In my first Chapter, I refer to The Oannes, an "Other" species that lived in the Persian Gulf and came ashore to teach in what is now modern-day Pakistan. They had the form of a fish but with the head of a man under a fish's head and the feet of a man under a fish's tail. Perhaps they were the occupiers of any Persian Gulf base, or moved in once it had been vacated.

ii) <u>India</u>

These sightings all come from the one book[26]. There is plenty of variation in "Other" species to assess:

- Giant Bird 1
- Yeti 7
- 1.7m 3
- 1.3m Hairy 4
- 1.0m 4

The giant bird could be an Avian but, as they do not appear to range widely, it is more probably one of the black monster variety. The number of Yeti is high, but they are close to where they are said to have originated. What is surprising is that it is claimed that two were caught in 1975. That is a very rare occurrence. The 1.7m hominid is difficult to identify without more information. The 1.3m hairy beast is probably a type of Dwarf, and the 1.0m creature could well be a type of Little People. There are apparently no Greys in this sample.

There are only 4 types of UAP represented:

- Cigar 2
- Disk 6
- Boat (?) 1
- Sphere 1

One of the disks was described as gigantic.

There is not the evidence to suggest that a base or a colony has being set up on the Indian Ocean coast.

c) PACIFIC OCEAN

i. Eastern Russia & the Sea of Okhorst

The Sea of Okhorst, Far Eastern Russia, and the Kamchatka Peninsula form a distinct unit. See Figure 2.

Back in the last ice-age, the combination of lower sea-levels and the expanding North American Ice -Sheet, resulted in a link between east Russia and west Alaska. Early humans were able to cross via this link, to populate North America, and possibly even travel further south.

Sightings in the table above come from 3 sources[27,28,29].

From the sightings reported so far, the Sea of Okhorst, seems particularly active, and certainly the Russians[30] believe there to be at least one "Other" base or colony in the area, possibly in Lake Kronotsky in Kamchatka. The largest UAP, a 200m cigar, was headed in this direction.

The local sightings of 2m "Others" suggest that the web-fingered Nordics, the Cassiopeians[31], have spread to this area from the inlakes where they were first reported.

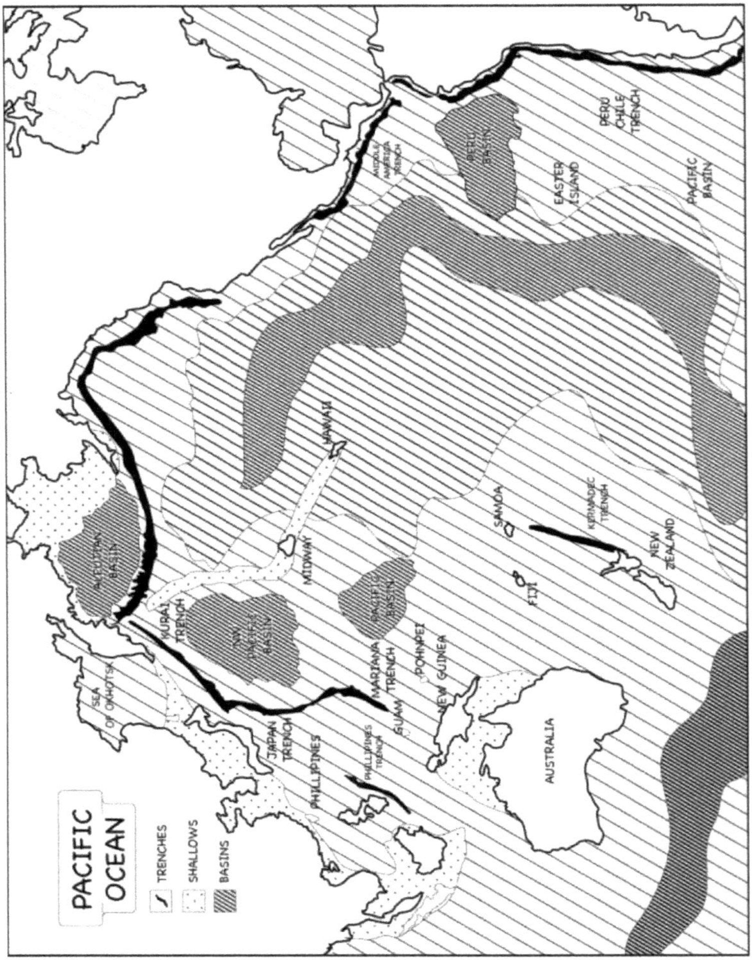

Figure 2 The seabed of the Pacific Ocean

ii. Coastal British Columbia, Washington, Oregon & Southern Alaska

It is necessary to divide Alaska into 2 segments to do it justice. The mainland is huge and seems to contain at least two hot-spots in the mountains, and the south eastern "Inside Passage" archipelago more comfortably belongs with the western lakes and islands of British Columbia, then down into Washington and Oregon.

All the Ketchigan sightings for British Columbia reported above come from a book by Debbie Zieglmeyer[32]. Someone there was keeping records and has passed them on, whilst presumably no-one was recording in the rest of the archipelago during the period from 2000 onward. Other sightings are from George Mitrovic[33].

Debbie also draws attention to the US Navy's SE Alaska Acoustic Measurement Facility (SEAFAC) which is just north of Ketchikan, and points out that there is a possible underwater anomaly in Prince William Sound[34], further north between Juneau and Anchorage. Is this another example of US and "Others" working together like in Dulce, the Bahamas and Pine Gap in Australia, or is SEAFAC just keeping an eye on them?

Considering the number of UAP sightings recorded in these states (35, plus clusters) there have been surprisingly few "Other" sightings. There were 7 disks, the preferred vehicle of the Small Greys, with 3 confirmed sightings of the creatures themselves, together with a Mantis which is always in charge of the Greys when present. The majority UAP, a sphere, was confirmed to have been observed at least 14 times, but this design cannot be associated with any one particular species of "Other", although Small Greys have shown a preference for disks.

Interestingly there were also sightings of a one-eyed giant, or Cyclops, and a Chupacabra, the voracious farm livestock eater.

iii. Coastal California

These records come primarily from three sources[35,36,37]. Certainly, following the nuclear detonations of 1945, we can see an increase in UAP activity in the south of California, and also in the very north, with multiple explosions and UAPs entering the waters around the Catalina Strait. It is likely that a new "Other" base was being constructed somewhere near Catalina Island.

The experience in California seems to fall into three phases. First of all, UAPs arrived around the time of the first atom bomb, and were seen over the whole State, then came the local explosions, combined with further UAP sightings and a few abductions. Finally, in the 1970s, things settled into what was almost routine, with more abductions and fewer reported sightings.

Decade by decade, reported abductions increased:

- 1940s 4
- 1950s 7
- 1960s 13
- 1970s 19
- 1980s 20
- 1990s 16

There are seldom going to be multiple witnesses to abductions, and little subsequent evidence unless there have been devices implanted, or the abductees are subjected to the questionable process of hypnotic regression. The problems with reported abductions are manifold:

- How many imagined it?
- How many elected to report it?

And most importantly:

- How competent was the hypnotist who helped them remember?

One of the distinctive features of many abductions is that the victim is told by the "Others" exactly what they want to hear:

- "Don't be afraid, I won't hurt you"
- "You've been selected for something special"
- "Something special is going to happen soon"
- "I have a message especially for you"

The victims often seem to have suppressed memories of Greys and Mantis-type "Others". This could be simply the spurious outcome of leading questions by a prejudiced hypnotist but, assuming it is not, the Small Greys seem to be running riot here, clearly indicating the probable presence of at least one Small Grey base nearby.

iv West Coast of Mexico & Central America

All these sightings come from the one book[38]. Here, there is one single-witness sighing which needs noting. In February 1995, a lorry-driver claimed to have found an "Other", 0.30m tall, sitting in his passenger seat. This individual said that he lived in a colony at the bottom of Lake Chapala, which is near to Guadalajara. It is a very big lake (75km x 25 km), but shallow, with a maximum depth of 9.0m. He claimed that he was descended from the occupants of an UAP that crashed thousands of years ago. This would certainly fit in with the small people described in Mexican folklore, that they called Chaneques.

There were some puzzling sightings in this table with, in one instance, a group of 18 human-height "Others" and, on another occasion, a group of 14 tall "Others". No-one felt inclined to stop them to ask what was going on, but it certainly seems strange even by UAP standards. If these two groups are excluded from any assessment, "Others" shorter than 1.0 metre high are the predominant entities, suggesting that there really is at least one "Little People" colony somewhere in the area. Given that they have also been observed on the other coast of Mexico as well, they seem to have made themselves at home.

v Peru

All these records come principally from the one book[39], and, I have also included a few aircraft sightings from the book by John Scott Chace.[40]

Sightings here seem to relate to population density, with many occurring in the Lima area or along the coastline. In particular, there is no evidence here of any preponderance between the coast and Lake Titicaca. Single-witness sightings claim there is an underwater base or colony there, but there is insufficient evidence to support this.

As usual, there is a mix of species contributing to the "Others" sighting list:

- Very Tall 2.2m 11
- Very Tall Copper hair 2
- Tall 2m fair hair 5
- 1.6m Egg shaped head 1
- 1.5m human-like 3
- Dwarf 1
- 1.0m Cyclops 10
- 0.8m greenish 8

As in Argentina, the list is dominated by 2.2m "Others", and very small "Others" of one sort or another. The sighting of a Dwarf could be misleading, because it could be one of the smaller ones. The one Grey sighting shows that they are, as ever, sniffing round for any opportunity.

The 0.8m species seems to be a bit frog-like and do not walk in the same manner other humanoids do, These could be the creatures known in local folklore as Karka, possibly an onomatopoeic name suggesting that they might have evolved on Earth and are what are known as Intra-Terrestrials.

The Cyclops-like "Others" have been described as having features which might be adaptions for a maritime existence, such as fin-like hands.

Unfortunately there is not the information to indicate where any "Other" bases or colonies might be. There is insufficient evidence to suggest one in or near Lake Titicaca, but there is an underwater trench running close to sea-shore which could, as in Argentina, provide a suitable location for a base or colony.

vi. Chile

Most of these records come from the one book[41]. Details of military and aircraft sightings come from another by John Scott Chace[42]. In Chile, there appear to be very few claims of abductions.

There are a few occasions where there has been a single sighting of a particular species of "Other". These are a Shape-Shifter, a Tall Hairy creature like a Bigfoot, what was probably a Chupacabra, a Cat-like creature and, of course the ubiquitous Small Grey. The more frequent sightings have been:

- Tall over 3m 9
- Over 2m no neck 1
- Over 2m Nordic 16
- 1.2 m 6
- Large Bird 4
- Less than 1m 6

The Nordics are often described as Angelic Beings. The Birds appear to frighten witnesses but do not appear hostile. Chilean folklore describes little people living in the forest, that they call Trauco. This may be historical proof of their presence in Chile for long enough to enter into folklore, something which doesn't happen overnight.

The dominating presences are the extremely tall "Others" and the Nordics.

There is no indication from the distribution of sightings along the coast that there is any sort of underwater base or colony. The most sightings are in the wider Santiago area, so perhaps there is an underground base inland there, accommodating the very tall whites and the Little People who are often seen together. Tere is an underwater trench off the coast of Chile which could also provide a venue for a base.

vii. <u>New Zealand</u>

All these records of sightings come from one book[43]. There seem to be far more sightings for these islands than for much larger countries. I suspect that this is due to diligence in recording rather than actual frequency.

Strangely there was a sighting of a large predatory cat, although New Zealand has no large cats amongst its fauna. Yeti-like creatures have also been sighted on several occasions. There were also a number of sightings of weird creatures in the sea, getting caught up in fishing nets. I suspect that the sighting of a giant 20m turtle is more likely to be the top of some form of UAP.

The early reports of airships, and then cigar shaped UAPs, all over North and South Island suggest a detailed reconnaissance of the islands during 1909. It would appear that the decision was made that this was a suitable place for a colony and, since then, one thing which stands out is the number sightings in or near Auckland on North Island. As Auckland lies at the south end of an undersea trench, it is certainly credible that an "Other" colony is now sited somewhere there.

Although "tall", "human-like", "dwarf" and "very small" entities have been sighted, there appear to be no records of Small Greys. There is insufficient evidence to deduce the "Other" species, with any certainty, but human-like entities are in a slight majority, so perhaps the earlier human civilisation, the Telosians, has set up home here again.

viii. Australia, Papua & New Guinea

The listing includes coastal sightings. Indeed, other than the Pine Gap facility near Alice Springs, there are very few sightings inland.

Most of these sightings are recorded in the one book,[44] with additional material from two others.[45,46] In this listing I have only included a few of the many sightings. I have listed all the UAP sightings in 1954, to show typical frequency in Australia. Sadly, in general, many of the incidents have only a single witness, so I would normally discount them. It is interested to note that many of the abduction events claimed by single witnesses have involved what are known as Small Greys.

There have been many sightings of strange creatures over the years. There were umpteen sightings of Hairy Monsters (Yowie), Waterpuppies which are perhaps freshwater seal, Bunyip, whatever they are[47], and Thylacines[48] such as the reputedly extinct Tasmanian Tiger. Single witnesses, whose reports I do not usually include, have claimed sightings of "Others" as small as 5cms in height.

Reports of creatures resembling Yowie have varied significantly in estimates of their height. Given that some family groups have been spotted, it would appear that there is a very healthy breeding population, with juveniles and females sometimes living apart.

There have also been multiple sightings of lions, tigers and panthers, although such big cats are not native to Australia. The question is how did they, and the Yowie, get there?

As you can see in Figure1, there was no land bridge to Asia in the last ice-age. Were some “Others” playing games? The tall hairy Yowie are still reported as present, but where have the others gone? Are they still there somewhere, did humans hunt them to extinction, or did the “Others” move them on? It must be remembered that there is reputed to be an “Other” species, the Urmahs, who are tall felines. They are generally bipedal, but that doesn’t mean they have to be if they are pretending to be something else. I haven’t heard of a basically canine species which could pretend to be Thylacine.

At sea there have been plenty of sighting of a large sea monster with a long neck, up to 20m long, and also of coastal sightings of metallic objects thought to be like submarines but in water too shallow for submarines.

There have also been several records of another monster, which has been given the scientific name “globster”, found dead on a beach. This has been described as having a 10m body and being almost 3m high, hairy, with 6mm skin then a layer of fat then solid meat. The interesting thing was that it appears boneless. Could it be an unknown relative of the giant squid? If you remember, back in Chapter 2, I suggested that a cephalopod might be the best bet for evolving into an intra-terrestrial.

The UAPs which have been reported are typical of those seen world-wide, with some exceptions:

- Large and small cigar shapes
- Rectangular bodies as big as tramcars
- Hexagonal
- Large double disks with portholes between
- Single disks below with a large dome on top
- Disks with a domed cockpit
- Egg shapes
- Spheres of varying size

In addition, there have been tiny metallic devices of various shapes reported as simply falling out of the sky.

Inland from Sidney lie the Blue Mountains. There are claims that there is an UAP base there. Typically, the area has been called the Blue Mountain Triangle.

Certainly, there have been some very unusual happenings there over recent years.[49] There are apparently 600 or so reports of unidentified flying objects that have been seen over, or on, the Blue Mountains. The Jamieson Valley and its surrounding environs has been the scene of many spectacular UFO events. Also, there have been repeated sightings of UFO's that seem to just appear as if out of nowhere in the almost impenetrable forest regions to the south of Mount Solitary and beyond the backwaters of Warragamba Dam.

However, the authors contend that this site contains a massive US/Australian base, bigger than Pine Gap, where the US manufacture their own spacecraft. They make no claim that "Others" are present in the base, but suggest that they may be in contact with them. Certainly there are enough sightings of "Other" craft in the area to suggest that the "Others" are keeping a close eye on them. I leave you to read their book, and to form your own judgement, but this doesn't explain where all the "Others", sighted in these areas, come from:

- Yowie 4
- Tall Human-like 2
- Tall Fair Nordics 4
- 1.8m no neck 2
- 4m lizards 1
- 1.4m red hair 3
- 1.0m dwarf 8
- 0.5m Little People 10

These figures suggest that the smaller "Others" predominate in Australia, along with the Yowie.

A Darlington[50], Perth, man says he's captured pictures of hundreds of UAPs from the veranda of his home. It began when he was taking photos of clouds to test out a new camera and he noticed a "smudge" that, when enlarged and enhanced, "had some structure to it, suggesting it could be some sort of craft in the sky". He says that, since then, he has identified a dozen different UAP types including round, square and saucer-shaped craft, posting the photos to his website wispyclouds.net for extra-terrestrial buffs and sceptics to ponder.

Overall, there are no positive indications of any maritime-based colonies in the area. They are either keeping their heads down, or are land-based.

ix. Malaysia and Indonesia

The pinwheels seem to be similar to the one sighted in the Gulf of Carpentaria in Australia and the multiple sightings in the Persian Gulf.

Beyond that, there appears to be no doubt that tiny "Others", Little People, about 7.5 cms tall, have set up a colony in Malaya, maybe in Terengganu on the eastern seaboard. There may also be a second species present, perhaps dwarfs, towering over the first at a full 1.0m.

a) Japan & the Dragon's Triangle

There were very few sightings of "Others", but I have been able to look at those UAPs where there was a description provided:

- Cigar shaped 6
- Disk 5
- Triangular 2
- Rectangle, Sphere, Egg, Cone 1 each

The island of Pohnpei in Micronesia contains the remarkable ruins of Nan Madol[51], which might prove to be the residual evidence of an underwater base. Also, in the mountains of Fukushima Prefecture[52] in Japan near Senganmori Mountain, the locals report frequent sightings of "Others" with wings and hawk-like features. These Avians sound similar to Anzu of ancient Sumeria or Horus of Egypt, and this may be their long-term Colony

xi. Coastal Russia

Most of the sightings recorded in this table come from three books.[53,54,55] The single sightings do give a pointer to abductions, claims of which seem to become far more common after the 1950s.

There are very few "Other" sightings in this list, with 6 tall, 2 Tall & hairy, two Dwarf-size, and one each of an "Other" with a tail, one with no neck and one winged "Other". It would seem that "Others" aren't too keen on being out in the cold either.

Amongst UAPs there is more of an indication:

- Sphere 14
- Disk 8
- Rectangle (all 12 in one sighting) 12
- Gigantic 6
- Triangle, Cigar, Cylinder, Bell 1 each

xii. Hawaii

All these sightings come from one book[56]. There does seem to be a wide range of "Others" described amongst these.

Menehune is the islander's name for a tiny "Other" less than 1.0m tall, which is known to play with children. In Mexico they are known as Chaneques.

It is clear that there were "Others" in Hawaii before the atom bomb, and that they have an underground base or colony which, based on the sightings, is likely to be in Oahu.

d) ATLANTIC, CARIBBEAN & BALTIC

i. Nova Scotia, Labrador & Newfoundland

These sightings come from 2 books[57,58].

There were only four sighting of "Others" in this area, three of them Little People, but there were several claims of abductions amongst the single-witness testimonies. The Atlantic is very deep between Greenland and Canada, and this was not covered by the North American Ice-Sheet during the last ice-age, so an "Other" base could be hidden there. Sea-ice would have been common, but the ice-sheet itself would not have reached out very far.

However, on the Quebec-Newfoundland border there is a mine, much deeper than others nearby, and which is guarded by the police, where Little People are reported to have been in residence for a long time – a Colony!

ii. Bermuda & the Southern East Coast of the USA

The Bermuda Triangle is not included in this area. It falls within the area covered by Puerto Rico below. Many of the sightings off the east coast of the USA are so far out to sea that their location cannot be specified in terms of their proximity to any particular place on the mainland coast.

There were very few sightings of "Others" in the whole of this area:

- Tall 2.2 1
- Nordics 3
- Large winged 1

- Greys 7
- Dwarfs 1
- Small 1

There were plenty of sightings of UAPs, including the UAP train on its way from Alaska to Brazil reported on elsewhere.

- Triangular 1
- Disk 12
- Bell 1
- Small Square 2
- Cigar 1
- Unspecified 24

These only give the hint that possibly the Small Greys have moved into the area, but no indication of where they have set up home. An unexplained explosion, heard over North Carolina, points to the fact that they probably have.

The closest other possible suspected UAP sites are in Newfoundland, the Bahamas and Puerto Rico, unless there is one within the Bermuda Triangle as claimed in several publications. However, there seems to be little or no evidence to support this.

iii. Gulf of Mexico & Its Coastline

The majority of the sightings recorded here come from three books.[59,60,61]

The underwater topography of the area is shown in Figure 3.

At least there are a few underwater sightings:

- The "Others" sighted by divers working on an oil-rig
- The two submersible UAPs
- The three long-necked sea monsters

Generally, the UAP sightings in the area are typical for North America:

• Disk	18
• Small Disk (1-2m)	7
• Sphere	12
• Egg-shaped	4
• Triangular	2
• Cigar	2
• Tadpole	1

As you might expect, the Small Greys preferred mode of transport tops the list, with spherical UAPs second. I'm never sure whether "Egg-Shaped" refers to the shape of the egg's shell, or as it comes out of the pan when fried. Both are credible. I suppose a "Tadpole" is a disk with a tail at the back.

There is some difference in the frequency of "Others", even though Small Greys still top the list, with so many US states bordering the Gulf:

• Small Grey	8
• 1.5m Tall	3
• Very Tall	2
• Tall & Hairy	2
• Tall	1

- Bat-Like 2
- Little People 2
- Lizards 1

Here the Little People are accused of abduction of young children, to the despair of their parents. However, in Mexico, they are known to look after them, feed them, play with them, and send them home after about 5 days.

They call them Chaneques. If they have been around long enough to pass into folklore, their presence is not modern. Their Colonies are probably underground or in lakes rather than under the Gulf.

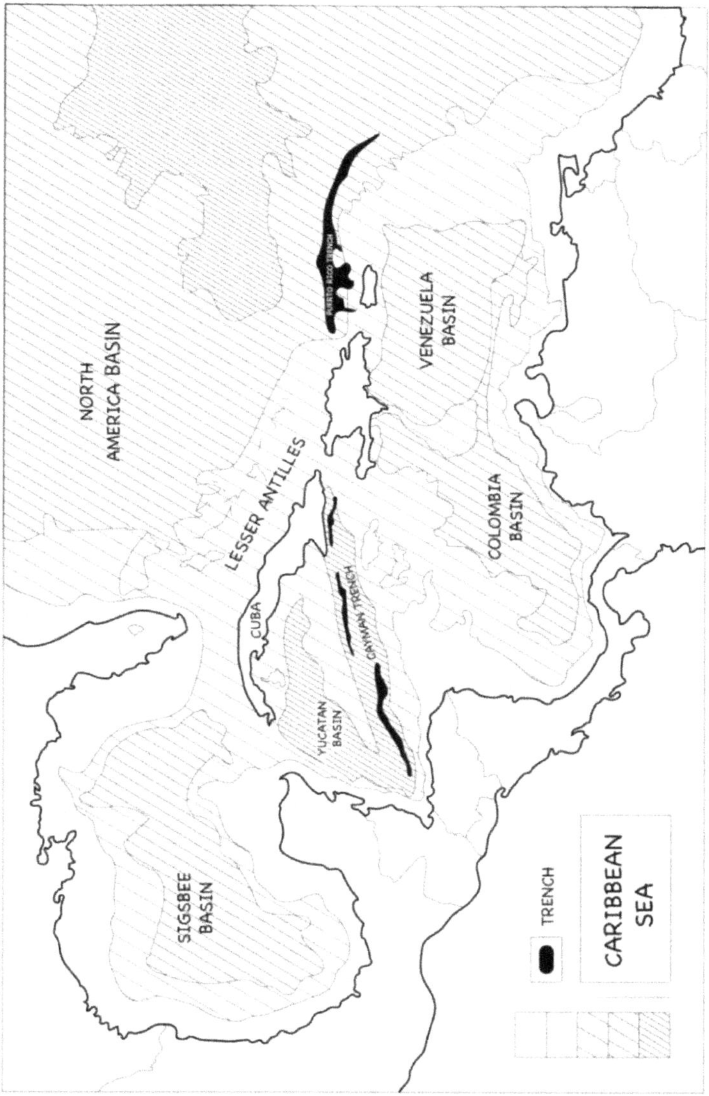

Figure 3 Topography of the seafloor of the Caribbean Sea

iv. Puerto Rico, Bermuda Triangle & South Caribbean

The majority of sightings in this table come from one book[62], together with additions from two others.[63,64] For its size, Puerto Rico is almost overflowing with various types of "Others". I have omitted dozens of sightings of spherical UAPs and of Small Greys. Nevertheless, many species are represented in the sightings:

• >2m tall thin	2
• 2m tall hairy	3
• 2m tall Nordic	10
• 1.3m greys with large heads and eyes	14
• 1.3m – 2m creatures with wings	6
• 1.3m greys with crest	1
• Dwarf	10
• Little people	10
• Canines	3
• 1.3m carnivores – Chupacabras	2

The predominant species appears to be Small Greys, even though they only started arriving in 1978, and there are repeated single-witness reports of their abducting people. They are regularly reported in the vicinity of Lake Cartagena and in the area of San Yungia, in both of which they are suspected of having bases. Both of these seem to be relatively recent, whilst the Colony reputed to be in the nearby Puerto Rico Trench, could be much older.

It is interesting to note that there is a disused gold mine in San Yungia. "Others" are well known for their interest in gold.

The next most frequent sightings are of Nordics, Dwarfs and Little People. The Little People are not known to like being underwater, and the Nordics have been around for longer than the Dwarfs so, on the balance of probabilities, the Nordics could occupy the Colony in the trench.

One of the islands of the Bahamas, Andros Island, hosts the US naval research base AUTEC. It is claimed[65] that there is an "Other" base in the deep water close nearby, linked by power and communications cables.

In Venezuela there were two attempted abductions by "hairy Dwarfs" who proved to be very strong, with apparently unbreakable bodies.

v) Cuba & West Caribbean

The only "Others" sighted are one Bigfoot and several large Winged creatures. It doesn't give any indication as to what the UAPs are doing on the island, although sightings of these could be used to support the assertion that there is an UAP Colony in Guantanamo Bay, even if it is not clear who is in residence.

vi. Brazil

Almost all these records come from one book by Thiago Luiz Tictti.[66] I have extracted records for the maritime provinces of Brazil. He does not claim that the list is complete, and he sometimes comments in a report that there have been other sightings in the same area, but he doesn't list them.

There have been many spectacular sightings in the rest of Brazil, but here I am specifically interested in underwater bases and colonies.

In my area of interest, "Other" sightings have been recorded as:

• Over 2m tall	14
• 1.9 m plus, tall and fair	5
• Human-like	6
• 1.3 m tall	33
• 1m tall	11

One tall reptilian has been reported.

Of these "Others" the group with heights about 1.3m and under predominates. These could be either the Dwarf-like "Others" or Small Greys, but there have been few claims of having seen the distinctive head and eyes of Small Greys.

Some of the reports of failed abductions describe the victim holding onto a tree, resulting in a protracted tug-of-war. This rough handling is more reminiscent of Reptilians rather than Small Greys with their mental abilities, although it would be a mistake to discount the tall fair Nordics who have been known to abduct humans.

There are fewer sightings of "Others" reported along the Brazilian coastline, than along the shorter Argentine coastline, but is much easier to remain unobserved in jungle.

Given the large number of UAPs reported in Brazil as a whole, there must be a number of UAP colonies or bases in the rainforest and in the Andes. There is reported to be one alien Colony in the ocean 100 miles south of Rio de Janeiro.[67] Tall "Others" have claimed to own this, but do they share it with any other species?

The sky-train event of 1913 describes a multitude of UAPs travelling line-ahead from at least Saskatchewan down the east coast of the United States, over Bermuda and down to the most easterly point of Brazil where they vanished, possibly into a newly created colony. What species were they? Reports from the surrounding states give no help.

vii. <u>Argentina</u>

In the Argentine as a whole, there are reports of at least two "Other" bases or colonies, one south of Cartagena and one off the coast, as well as a wide variety of "Others". Over the country as a whole:

- The majority presence is of tall thin entities well over 2m tall. Also at 2m, is a broader hairy version described as being like a Yeti.
- There is also a Tall entity with a small head so that it appears that its arms come out of its neck.
- There are also sightings of tall lizard people, always underground.
- At typical human height is an orange skinned entity. Also a beautiful blond lady is described who warns of floods, and hovers above water.

- There are reports of 1.5m tall entities with large heads, big eyes and often wearing a light on their head.
- The 1.2m tall creatures are often described as gnomes or dwarfs, and seem to have green skin.
- The second most common species is about 60 cm tall - the Little People

In the following table, I have confined myself to reporting only sightings of "Others" in Argentina's four coastal provinces. All these records come from one source[68], although that one is prone to duplicate records.

The presence of various "Others" here is just as diverse:

- The Little People 45
- Very Tall 37
- Nordics 20
- Dwarfs 20
- Small Greys 19
- Felines 6
- Lizards 5
- Tall Orange Skins 3
- Yeti, No necks, Mothman, Canines 1 each

It can be seen that, inland, there are multiple species of "Others", but on the coast, there are more sightings of very Tall "Others" – Arcturians - and the very small 0.7m Little People. Perhaps they work together. There was some form of a major upset in 1968, when there was a multiplicity of sightings of very Tall "Others", but it is not evident what the cause was.

In addition to any inland bases in Argentina, there is reported to be an underwater colony off-shore at Mar del Plata[69], a few hundred miles south of Buenos Aires. This appears to be the domain of the very Tall "Others". This consistency tends to confirm that coastal Argentina at least has been colonized by very Tall "Others", perhaps together with the fairy-like Little People.

viii. <u>South Africa</u>
These sightings all come from the one book[70]. Interestingly there are many sightings of sea serpents also recorded, stretching into the 1950s. I have omitted these.

There was a flurry of UAP sightings in the Eastern Cape in 1972. Prior to that, in 1914, a single UAP was reported to have over-flown a lot of the country. Was this reconnaissance prior to establishing a colony in the Eastern Cape? The presence of a group of 20 one meter tall "Others" near Durban in 1986 suggests that the Little People might be in occupancy.

ix. <u>Portugal, the Azores and the Canaries</u>
It is possible to produce a narrative based on these sightings, provided they are accurate:

> *There was an "Other" base at first in Tenerife until one of their tunnels was found there in 1912 and they moved to the mainland. They remained there in the Algarve, until their underground tunnels there were found in 1962, and whoever was in occupancy felt they had to move again. Their base or colony is now back in the Canary Isles but presumably much better hidden.*

However, it would only take one error to change this. If the tunnel found on Tenerife was really a lava tube, we end up with the initial base being in the Algarve. I suppose that the same could be said about the tunnels in the Algarve, but it is evident that something significant happened to sightings in Tenerife in the 1960s, with none recorded before then. Either a base or colony was moved from the Algarve, or a colony was set up from scratch.

Some massive UAPs have been sighted in the area during that period and afterwards, so the colony would have to be pretty well concealed, probably with an underwater entrance because, after these sightings, the local inhabitants would have had plenty of incentive to go looking.

Too many different species of "Others" have been sighted since then to determine which is in residence, or whether it is the classical HMO – House of Multiple Occupancy.

x. Norway

Records of strange happenings in Norway[71] go back to the 1700s. Many of these describe sea-serpents, lake-monsters, the disappearance of children, stories of strange creatures, and attempts to lure young women away. This is so much part of Norwegian folklore that it cannot be taken as definitive evidence of "Other" activity but, if a colony has been there long enough, it may well become part of traditional fables. Many stories stand out as offering this possibility.

It is difficult to tell whether there is an "Other" colony in Norway, where it might be, and what species of "Other" might occupy it.

The best bet, on the basis of the multiple UAP sighting over many years is Hessdalen near Trondheim. This could either be an underwater colony or one in a mountain top. There appear to be multiple sightings of small "Others", particularly at Honefoss, near Oslo, but also riding within UAPs in many other sightings.

Adding to these considerations are the events which have occurred in next-door Sweden:

- In 1946 there were over 50 sightings of 3 - 4m long cigar shaped UAPs which seemed to be deliberately crashing into freshwater lakes
- From there on, there have been multiple sightings of flying saucers of the sort familiar to North Americans.

It would seem that Norway, combined with Sweden, is densely colonised by "Others".

xi. <u>Sweden</u>

Sweden appears to have a major infestation of "Others.

This commenced long before the Second World War, and may be a prime example of "Other" colonisation of the whole of Scandinavia, with the colonists initially trying not to interfere in our Human society. The Atom Bomb changed this.

However, for the purposes of this book, which is concentrating on "Others" in our waters, the arrival of the cigar-shaped UAPs in the Swedish lakes is my primary interest, along with the possibility of a colony in the Baltic or a Swedish lake,

There are records of the arrival of 6 of these so-called "Ghost Fliers" in 1934, followed by at least 50 in 1946. This is not to say that nothing happened in between, but we only have a few records or sightings during war time. Did the Ghost Fliers continue, or did the war disrupt them? We don't even know what type of "Other" to look for.

Overall, there have been sightings of entities less than 1m tall, beautiful human-like entities that look like teenagers, adult size humans that seem a bit strange, very tall entities and Yeti-like entities.

There have also been reports of strange creatures in lakes, going back years - Sweden's "Lake Monsters". One major example of this is Lake Storsjon, joined to the sea, where the first reported sighting was in 1820. It was described as having a body 8 – 10 metres long, the width of a rowing-boat, with a 10 metre long neck almost as thick as a man's body and with a head like a red-grey horse with a white mane. The last reported sighting was in 1965.

The distribution of sightings across Sweden is fairly uniform, if reckoned in terms of land area, but not if considered in terms of population.

The whole country has 29 sightings, whilst Lapland has 10:

	Country	Lapland
• Disk	16	10
• Sphere	4	
• Cigar	4	
• Oblong	3	
• Pyramid	1	
• Triangle	1	

This gives a total of 29 UAPs, one third of which are in the least inhabited areas. The same statistic applies to disk-shaped UAPs if considered alone. This suggests that something different is going on there, and that we should look for a possible "Other" base.

When we look at "Other" sightings, a different picture arises:

- Dwarf 11
- Small Grey 7
- Cyclops 2
- Slugs 2
- Tall Thin 2
- Little People 2
- Tall Black Bird 1
- Hairy 2m 1

There is only one "Other" sighting in the whole of Lapland.

In practice, it is probably difficult to differentiate between the Dwarfs and the Little People in these sightings, so it may pay to count them as one.

It should be noted that the King of Sweden was visited unannounced by 3 Dwarfs in 1941. It may be that they were declaring their presence and seeking permission to set up home there.

So what we could have is a colony of Dwarfs in northern or southern Sweden, with some Small Greys and a smattering of the other common species. There may well be a separate colony, perhaps with a different species in residence.

Unfortunately, we have no idea what was carried in all those Ghost Fliers. They were Little People sized, but were they land-bound, amphibious or a water-breathing species?

e) INLAND WATERS

i. Russian Lakes & Rivers

These records come from three sources[72,73,74]. There are many recorded observations in the Ukraine which have involved sightings of "Others". There have been some which are probably Small Greys in their flying saucers which are familiar to North Americans. However, the majority have been of spherical UAPs carrying varying types of very tall "Others". There have been 3m humanoids looking like enlarged modern human kind, and it has been suggested that these are descended from the Titans of ancient lore. There have also been similar entities with much smaller heads, and very hairy versions described as like a Yeti.

Some have claimed to have seen webbed feet and hands. However, there is the possibility of these being part of their immersion kit which also includes a spherical helmet as described in the Lake Baikal incident in 1982.

These spherical UAPs are recorded across the whole of Russia/USSR, although there are fewer on the coastlines where the proportion of flying disk increases. This implies that Tall "Others" are also widely spread over inland Asia. It is not clear whether they are showing a preference for fresh water over salt water. Sightings in the Baltic and Black Sea suggest they don't have a total aversion the salt water.

Looking at the large number of single-witness sightings relating to abductions, it would seem that, in Russia, the role of Small

Greys in the USA has been taken over by the Tall "Others, undertaking much the same procedures.

Given the multiple sightings in Lake Baikal, there could be a colony there, but they could well have a second, western site somewhere.

ii. The Great Lakes

It is suggested that there are at least two UAP bases in the Great Lakes.

Lake Superior is 1,333 feet deep, Lake Michigan 923 feet, Lake Huron 750 feet, and Lake Ontario 802 feet. These would not be deep enough to hide a base from current detection methods, so any UAP bases would need to be buried beneath the lake bottoms.

There are no sightings of "Other" entities reported on the lakes, so we need to look for any abductions reported in the surrounding areas if we are to identify the species occupying any bases there.

Lake Michigan is the location of one of the claimed Mystery Triangles, and this one includes, along with various UAP sightings, disappearances of both ships and aircraft, with some leaving no trace.

It may be that the Great Lakes are purely an UAP refueling stop. They seem to require water as part of their propulsion system, so they may simply using one of the best sources of fresh water in a stop-over before heading for their Earth base somewhere else, or outwards.

iii. Northern Italy

Mount Musinè[75] in northwest Italy near Bonere has had a reputation for strange goings-on right back into Roman times. There are also nearby petroglyphs that are claimed to show "Others" and UAPs. There have been fewer sightings there of late, which might suggest that the base may have been moved.

During the early years of the 1800s, there were several reports of an explosion followed by an earthquake in Piedmont, and of various materials such as red soil, coal and yellow rain falling from the sky. This was accompanied by increased UAP activity at Mount Musinè. The explosions seemed to be centred under the Apennine Mountains. It has been said that the explosions caused the earthquakes, and that someone was excavating a new base there. Whatever the excavation technique may have been, the material which was removed could have been ground up and ejected into the atmosphere to fall where it will.

These references almost all come from the one book[76], but it contains many single-witness reports, which I have chosen to discount. There is a more specific listing,[77] but unfortunately, this only catalogues sightings up to 1978. There is reputed to be a second volume continuing up to 2020, but this is currently unavailable. In the meantime, I have added a few extra more modern sightings from another book.[78]

It is interesting to see the distribution of "Other" species in the area:

- Over 2.2m humanoid 5
- 2.2m Bat-like 2

- 2m Lizard upright 2
- 2m no neck 2
- 2.2m hairy 2
- 1.7m Greys Big Head 7
- 1.3m Helmeted Dwarfs 15
- 1.0m 9

It appears that the smaller species of “Others” are in the majority with the Small Greys pushing in. These are also present in many of the single-witness abductions.

Looking at the distribution of “Other” sightings around the provinces, it is not easy to identify the presence of any bases or colonies. There are sightings of disks emerging and entering the Adriatic in the Marches province, but the terrain there does not really stand out as suitable for a possible underground or undersea base. On the other coast the province of Liguria, which has more sightings for its size than might be expected, has predominately orange spheres in its area. Here the terrain, dominated by the Apennines, could suit an underground colony with an underwater exit.

Perhaps there are now two sites in northern Italy, one on each coast.

There is a significant presence of Dwarfs in the area, particularly in Liguria. It is certainly possible that Dwarfs have an underground colony with an underwater entrance near Genoa.

Chapter 4

WHAT "OTHERS" DO WE KNOW ABOUT?

a) GENERAL

As I have already mentioned, there are four possibilities for the origins of "Others":

- A more advanced version of the Human Race (Telosians)
- Species which evolved on earth faster or earlier than humans (ITs)
- Species from parallel universes (PUEs)
- Species from off-Earth (ETs)

The presence of one of these does not preclude the presence of any or all of the rest.

I have been trying to obtain realistic images or descriptions of some of these "Others" but the various publications on the subject are contradictory and, on some occasions, quite fanciful.

It is clear from many of the sighting reports, that it is dangerous to get too close to UAPs. They appear to be either dangerously radio-active, or emitting dangerous levels of Electro-Magnetic radiation, leaving humans with burns on their skin or, at worst, cancer and even death afterwards. It is ironic that many "Other" races are reported to be concerned about our use of atomic power, yet they are cavalier about the effects their technology has on us.

It would be a grave mistake to assume that these colonist "Others" only wish to live in peace with us. It would appear that the Galaxy is a dangerous place, and the various "Other" species are often playing a deadly game of one-upmanship. To some, we are, at present, simply a possible food source, whilst others are trying to protect us from their excesses. Unfortunately, some of these predators are colonists on Earth.

I will deal with the different types of colonists in groups as appropriate. I am going to use the names given to them by the authors of various books on the subject.[79]

My main difficulty is that some of these sightings sound like fantasies, but they have been witnessed by at least 2 people, so I must treat them as credible.

b) WATER MONSTERS

Right the way into the 1950s, there have been reports of monsters of the sea and lakes and, whilst it was fashionable to ridicule such sightings as a fancy of the imagination of early sailors, this accusation cannot be laid at the door of more modern sailors. Sightings from the 1850s onward have included:

- Norwegian lake-monsters, whatever they may be.
- Sea-serpents about 20m or longer. These can be very aggressive. They could have evolved from a creature such as the oarfish.
- Long necked beasts with flippers and a tail. The neck may be typically 5m long. These are apparently like the supposedly mythical Loch Ness Monster.
- Boneless “Globsters” whose bodies have only been found when dead or dying. These could be Cephalopods - an evolved squid of octopus.

These could all be evolved versions of Earthly creatures, and could match or exceed us in intelligence. Remember though, that Cephalopods may not be originally from earth, although their arrival was so long ago that this distinction is rather moot.

They need not necessarily have bases or colonies as we understand them. They would, after all, be living in their natural environment.

None of these could be mistaken for the early water gods which are honoured for teaching our early ancestors fundamental technology. Mostly, those were at least partly human in appearance.

c) GIANT APES

These occur all over the globe, may have local names such as Sasquash or Yowie, and may differ in colour, but they are essentially the same species. Males are typically 2.2 metres tall with smaller females.

They have been observed in small family groups of male, female and child, when they tend to avoid confrontation. Single males can be much more territorial, as can what are probably smaller juveniles.

Humans are reported to have first met these creatures in Tibet, where the received the name Yeti[80]. This is the title I will use rather than any of the ones that have been used elsewhere in the world since then.

If they originated in the mountains of Tibet, they could have reached the Americas in the last ice-age when there was a land-bridge between north-east Asia and Alaska, thousands of years ago.

Now we come to the interesting question: How do they range so far? They have been observed in Australia and New Zealand, and there was no land bridge to Asia during the last ice-age. Also, they have been reported to use weapons to immobilise humans, although this is not common. They are sometimes reported as wearing some type of medallion with technical capabilities.

The proof is not there, as we have no photographic evidence, but it is extremely likely that Yeti are Extra Terrestrials and are far more intelligent than we give them credit for. They live an idyllic life, which may actually be a life-style choice. They also have the intelligence to avoid the human hunters who want the fame for killing one, and have never left a carcass for someone to use as a trophy.

They may have access to the necessary technology to fly around Earth, or perhaps are able to call on help from other ETs.

They live off the land and are not hostile to humans. Our ideal neighbours.

d) THE GREYS

In practice there are at least three species of Grey "Others",[81] two of whom work together, sometimes doing very nasty things to live human abductees.

- Good Greys

Zeta Humans, from the constellation Cetus, are from 1.0m to 1.5m tall. They have thin bodies, bulbous heads, long arms, three fingers and eyes twice the size of ours. They are described as joyful and intelligent.

- Small Greys

Small Greys, originally from the constellation Reticulum, are from 1.3m to 1.6m tall. They have thin bodies, bulbous heads, long arms, three fingers and very large eyes which wrap around their heads. They practice genetic manipulation and reproductive system surgery on their abductees. They are reputed to have no empathy or remorse for their victims. At first glance, it is difficult to differentiate them from the Good Greys.

- Tall Greys

Tall Greys are also originally from Reticulum, are 2.0m to 2.5m tall and also have bulbous heads, thin bodies, and giant insect eyes which are twice the size of humans.

e) <u>TALL WHITES</u>

There are four types of tall whites which have cropped up in the preceding analysis;

- The Cyclops[82]

These are described as having a single eye and tall at 2m, but only one of the "Others" sightings was that height. The remaining 10 were much shorter. Were they juveniles or are these a separate race? They were described as being well adapted for swimming.

- Reds

These are described as being 2m tall with red or orange hair and orange skin. Nothing is recorded on what they do.

- No Necks[83]

These are again about 2m tall, but they have no apparent neck, so that it looks as if their arms come out of their head. They are occasionally seen moving around but there is little to describe what they are doing.

- Over 2.5m Whites

These are the tallest of the whites which have been recorded here, and are describes as of slender build. They are termed Arcturians in some of the publications,[84] on the subject of extra-terrestrial species. They are generally known as one of the "Good Guys", working to improve mankind's lot. Internationally they appear in sightings about as often as the Small Greys.

In speaking with contactees, the Arcturians have stated that they have colonies in or near the underwater trench which runs up the east side of South America although, considering their prevalence, they must work from other locations as well.

f) NORDICS

This species is the most ambiguous observed. They look almost angelic, are reported to render assistance to humans in trouble, but it is claimed that in Russia they have undertaken abductions in the same manner as Greys, but not as frequently.

However, there is more than one Nordic-looking species. The Cassiopeians[85] are the one Nordic species which has webbed fingers. They have been described as a benevolent species, but we may be mistaken. Perhaps it is they who are at work in Russia, using all the lakes there as homes. They may have a Colony in Lake Baikal.

The most likely benevolent Nordic species is the Apunians[86], who would certainly act to assist humans where necessary.

They are recorded as having met with President Eisenhower 1954 at Edwards Airforce Base to offer technology in return for nuclear disarmament. Ike refused, because he would not do this unilaterally. They ended by recommending he did not deal with another species called the Small Greys. Unfortunately he did.

They are over 2.0m tall, with fair hair and a light complexion, making them look Scandinavian. Hence the nickname.

It is likely that they have a long-standing Colony in the trench to the north of Haiti, based on the frequency of sightings nearby.

g) DWARFS & LITTLE PEOPLE

There have been sightings of Dwarfs[87] all over the world, and they vary greatly in skin colour and facial features. In general they are between 1.0m and 1.3m in height, and can be very hairy. They are extremely strong and, according to those who have got in a fight with one, virtually indestructible.

Dwarfs have become part of our folklore, which shows that they have been here a very long time. They were openly spoken about in Roman times, as individuals if not as a race.

Green skinned Dwarfs have given rise to the standard term for "Others" – Little Green Men – and the standard name, Martians. There is no evidence to suggest that they actually are Martians.

In the Middle Ages, they tended to fade out of our consciousness, although there were always rumours of hidden Dwarf races in distant lands. "Beyond the Atlas Mountains" was a favourite.

It seems likely that, in modern times, the Dwarfs originally had a colony in Mount Musinè in Northern Italy, and that they have re-located to the Genoa area, perhaps because they could no longer remain hidden where they were.

It is doubtful whether this is their only colony, but their others could be inland in obscure mountains where their secret presence is not under threat.

Their construction technique seems to include very loud explosions, followed by earthquakes. If this happens elsewhere, such as the Catalina Strait or the Eastern United States, this is an indication that they are at work again, even though it may be on behalf of another species.

Smaller than the Dwarfs are the Little People, ranging from about 1.0m tall right down to about 0.15m. Their tiny size has led to their absorption into native folklore around the world, under whatever name they are given locally. Again, they must have been here for a long time.

The Little People are not all one race. In Mexico, where they are known as Chaneques, they are known to steal children to play with them, to look after them for a few days, and then to return them safe and well fed.

There, it is likely that they have at least one colony, claimed to be under a shallow but vast lake near Guadalajara called Lake Chapala. As they have been sighted on both coasts of Mexico, they probably have at least two sites somewhere.

In Mexico's case, the Little People seem relatively harmless, but around the world in Malayasia, things are different. They appear to have arrived in the 1960s and set up home underground, perhaps under Terengganu. The Little People are only 7 centimetres tall and are very aggressive. If threatened, they are quite prepared to fight and to do harm.

In Java, a local man claims to have captured some of these people and exhibits them. He states that, to keep them healthy he has to feed them human blood every month. Apparently, they are not particularly pretty to look at, having protruding eyes – not the traditional fairy image.

It is quite likely that there are Little People colonies all over the world.

h) WINGED CREATURES

There seem to be three different types reported in the previous chapter. There is the black giant that looks like Dracula, Avians, and the dwarf sized sort that looks like an ugly Cherub.

The black giant has been seen worldwide, although it is a loner, and may even have achieved notoriety in the USA as "Mothman"[88]. There is no doubt that this creature scares everyone who sees it, but I have found no record of it hurting a human. It may be different for any animals when it gets hungry. It has been suggested that this creature is a Parallel Universe Entity (PUE), though I do not know how to prove this either way.

The Avians[89] are an ancient race that originally appeared in Sumeria, after the last ice-age working with the Anunnaki. They have hawk-like heads and wings, but it is not clear whether they can fly. There was one there called Anzu who was considered a god. Later in Egypt there was a similar being called Horus who was also considered a god. The only recent record comes from Japan in Fukushima prefecture, where they are believed to have a colony in or near Senganmori Mountain.

The ugly winged cherub has been harder to locate. The one possibility which I have discovered is that there was perhaps a case of partial miss-identification. There is reported to be a species of Dwarf with extremely large ears[90]. These might have been mistaken for wings in the heat of the moment. These Dwarfs are reputed to be able to fly around using their technology, not wings. They have been reported mainly in South America, but not in sufficient numbers to suggest that they have a base or colony there.

i) LIZARD PEOPLE

There has only been the occasional reported sighting of lizard people in my analysis, but they do have a very mixed reputation. They live underground, and one nest was discovered in 1870 underneath Buenos Aires[91].

There are reported to be many species of lizard people, and it has been claimed that one species is a very ancient earth race. Be that as it may, the one we have been seeing in Argentina is a predator on humans, which it abducts. There have been reports on failed abductions where the victim has survived by hanging on firmly to a tree trunk. Lizard abductions lack the finesse of a Grey's abduction because they lack their mental control capabilities. As they seldom leave survivors once they have made their capture, it is difficult to determine the rate of attrition of humans.

j) CATS & DOGS

There are some tales of feline looking creatures turning up in unexpected places, and more tales of canine creatures around the world.

In both cases, there are reported to be "Other" species which could be mistaken for these, particularly if they were deliberately avoiding identification.

Various big cats have been reported in Australia, where they are not thought to be natural fauna. It has been suggested that they are really an "Other" species called the Urmahs[92]. These are naturally bipedal, but there is no reason for them to be so, when they are pretending to be something else.

There are also reports of a canine in Australia called Thylacines, known locally, because of its stripes, as a Tasmanian Tiger. This was thought to be extinct. There have been reports of "Other" canines in other parts of the world, such as Haiti, but there is too little information to pin them down. Some writers claim that they are called Canis, and come from the Dog Star in the constellation Canis Major, but that does seem a bit too convenient. There was an Egyptian god with the head of a dog, called Anubis. Given that other Egyptian gods are reputed to be Extra-Terrestrial in origin, could it be that this applies to their dog god too? There was an almost identical dog god in Mexico called Xolotl. Both were gods of the underworld.

k) CHUPACABRA

This is another species whose presence is ambiguous.[93] The basic question is whether they are "Others" or escaped "Others" pets. They are also known as Goatsuckers, which gives an idea of their propensities. On the negative side, they are known to attack and kill domestic animals, to drink their blood, and to eat parts. On the positive side they are not known to have attacked humans. They are most common in the Caribbean, Brazil and Florida.

l) TELOSIANS

In my previous book[94], I suggested that there was possibly an earlier human civilisation, say 20 thousand years in advance of us, which had already left Earth and now just maintained a few bases here. These are sometimes called Telosians.[95] They look identical to present day humans

m) A WARNING FOR RESEARCHERS

Different publications give "Others" different names, assert that they come from different star systems, and ascribe them with different attributes. Some authors seem to look on the different species as a creative artistic challenge. It is convenient to be able to have names for them, but it would help if they were consistent. In the meantime, it is hard to know who to believe. The only ones on which there is any agreement are Small Greys, Nordics and very Tall Whites, but there is not enough agreement to decide where they come from.

Despite their presence worldwide, these authors mostly avoid discussing Dwarfs or the Little People. Some have gone so far as to describe them as magical creatures which must be left alone. They do this despite their showing very physical capabilities and responses to aggression.

In the words of Arthur C Clarke, "Any sufficiently advanced technology is indistinguishable from magic".

Chapter 5

WHERE ARE THEY HIDING?

a) BASES v COLONIES

In this book, I have concentrated on the underwater presence of "Others", and especially on those who need to construct facilities to live there.

"Others" have constructed many bases, but not all of these could be considered to be colonies. The particular requirements of a Colony are:

- They must be large enough to enable the inhabitants to continue the way of life they led on their home planet. Sometimes this requirement may be incompatible with an underwater structure where space may be severely limited, and even with subterranean caverns. It is understandable that "Others" seen to roam all over the surface even when they would be better advised not to.

- They must be well hidden, both from the local inhabitants, and from any malevolent "Others" who might visit the planet. Not surprisingly, "Others" exhibit what we would call "Human Nature" and do things for enjoyment even when they would be far better advised to stay concealed.
- They must be self-sufficient so that, if their home planet is destroyed, they can continue to preserve their species. It must be difficult to take such a long-term perspective when, hopefully, "It will never happen". Physical self-sufficiency is one thing, but social isolation could be very different.

In comparison, a Base can be much smaller, only needing to provide the facilities needed for whatever activities are being undertaken there. In addition to the military base(s) needed by the Small Greys to defend this planet, "Others" have bases at mining sites, and perhaps out-stations so that they can keep an eye on their neighbours. Trust is a potentially expensive luxury in this galaxy.

Inevitably, there are going to be more bases than Colonies. One author[96] has estimated that there are almost 50 "Other" bases in the United States alone and, for some reason, this doesn't include Alaska, where we know of at least two major facilities. We only know of two bases in the main part of the US that are Colony-sized amongst the 50 – Dulce and Catalina, although they may be more.

Dulce base is claimed to house about 18,000 "Others", reportedly Lizard People and Small Greys, and some of the activities they are reputed to undertake there would be considered totally unacceptable if we were in any position to prevent them. The Catalina base appears to be more military having, on one occasion, launched 200 disks all at once for reasons unknown. It did sound like a World War Two fighter scramble.

At least, when you are working at one of the smaller remote bases, you have the prospect of going "home" to your Colony,

These suggested motivations, of course, assume that an "Other's" emotions and thought processes are anything like ours. We can understand selfishness, as that is a trait we hold in common with at least some of them, but it is harder to understand altruism or service for the common good.

b) <u>THE COLONIAL ATTITUDE</u>

If you think back to our earthly colonial times, the attitude of the colonists then to the previous inhabitants of the lands they took, left a lot to be desired:

- The Spanish conquistadors stripped South American states of all their gold, and wiped out some tribes.
- The English financed its industrial revolution by robbing India of all its riches.
- The American settlers took the land away from the Native Americans, and even attempted to wipe them out by giving them blankets infected with smallpox.
- The Arabs looked on the Africans as potential slaves, selling them to the Europeans.

In general we looked down on the natives, describing them as uneducated and godless, and believing them to be at best greatly inferior. We could respect their courage at war, but didn't hesitate to kill them in their thousands with our superior weaponry. Massacres of them were considered just, but massacres by them were moral outrages. We even bombed their defenceless villages from the air.

We have to hope that our ET colonists have a more enlightened attitude towards us.

c) THE COLONIES

There are no indications of colonies which precede the last ice-age, although there may well have been a number of "Other" bases then. The Anunnaki would have to have somewhere to work from whilst tweaking our DNA, and that of any more earthly species that they fancied. After the ice-age, the water-based early gods which I've described, the Oannes, Nommo and Dagon, could have been there in Colonies, in the seas, but we don't know where. They could possibly have been in the Mediterranean.

Over all, following the ice-age, there were a number of "Others" (sky-gods) who are described as descending from the heavens, and taking up residence for a period to teach the locals. There are many impressive ruins from that period.

The first arrivals in more modern times were probably the Dwarfs and the Little People, who have been here so long that they have entered into our mythology. This is what makes it so hard for us to accept that they are truly "Others"

- The Little People likely have a Colony under Lake Chapala in Mexico, having survived a crash there. They probably have a second Colony in Mexico by now. There is another Colony or Base in Malaysia, one in Sumatra and one in South Africa. In practice, as the Little People sightings are so widely spread, they may have set up Colonies in a number of places. Their small stature makes them easy to hide.
- Dwarfs appear widely in European folklore, and are represented in several nations' pantheons of gods. They are traditionally famed for their blacksmithing skills. In modern times, there have been fewer reports of Dwarfs, although they do seem to have set up a Colony in Northern Italy near Genoa, with an underwater entrance, and perhaps a Base on the other coast near Marches. They are reported to have set up home in Sweden, perhaps on the Baltic coast.
- The Small Greys first set up a Colony in Sweden but, following the atom bomb and their "Treaty" with the USA, they have also set up at least two Bases there, large enough to be called Colonies. They have expanded down into the Caribbean and across into Europe.
- The web-fingered Nordics have a Colony in Lake Baikal in Russia, and have spread widely from lake to lake as far as the Baltic and the Black Sea in the west and the Sea of Okhorsk in the east.
- If the other Nordics have a Colony, it is most probably in the underwater trench near Costa Rica.

- The very Tall Whites, the Arcturians, have a Colony on the east coast of South America, perhaps in one of their two underwater bases near Mar del Plata and Rio de Janeiro in the deep trench there. A new base further north was set up in 1913, and they may be there too.

There are two Bases which may be somewhat different. In Hawaii and the Canary Isles, these could well be multiple-occupancy, given the large variety of "Others" species sighted in such a confined area. Could they be some form of oceanic regulator?

Chapter 6

CONCLUSIONS

I have shown that there is every probability that there are at least nine "Other" Colonies present on Earth near the coast, with four of these having underwater access.

The difficulty I've had in identifying Colonies is that we cannot see underwater structures. I have been able to identify concentrations of particular species of "Others", and to hypothesise on the location of their Colony, but cannot prove it.

However, often it has proved possible to demonstrate that the Colony is close to the coast, even if it is on land. As many of the Colonies are fairly new, this might suggest that creating an underwater Colony might be the ultimate objective, but a subterranean one is quicker to produce, and does provide an interim solution.

Some "Others" may never wish to go underground, yet alone underwater. Any Avian might find it too claustrophobic, and prefer an open mountaintop. On the other hand, any remaining Telosians have themselves well-established in mountain and Antarctic caverns, and in underwater structures.

Ultimately, the world's armed forces may well have detailed knowledge of the whereabouts of more Colonies, but governments are still under the misguided belief that they must not tell us mere mortals. Most probably, it suits the Small Greys to leave things the way they are, and our governments are doing what they're told.

Based on this work, UAP investigators will need to further address those questions that I have only begun to tackle. Who are our colonists and where are their colonies?

There will be occasions when the Colonists, in their drive for self-sufficiency, want the same resources that we do. There are certainly mining bases out there already, although we have yet to harvest the sea-bed for all its minerals, as they are capable of doing. At present they are simply helping themselves, but what happens when, inevitably, conflict arises?

Then we need to consider the most vital one: Can we evict those "Others" that are acting against our best interests?

As a world, we are facing many challenges in dealing with the "Others" whether they are just visiting or colonising our planet. Are our governments dealing with them in the best interests of the population as a whole, or simply in the interests of our governing elite?

We need to find out – and quickly!

APPENDIX

Tables of sightings

PERSIAN GULF & RED SEA

Date	Place	Credibility	Details
8th May 1879	Persian Gulf	Solid	Pinwheel
May 1880	Persian Gulf	Strong	Pinwheel
5th Jun 1880	Arabian Sea	Solid	Pinwheel
4th Apr 1901	Persian Gulf	Moderate	Pinwheel
26th Aug 1902	Nr Gulf of Aden	Solid	Pinwheel
1906	Gulf of Oman	Moderate	Pinwheel
17th Feb 1912	Off Pakistan	Solid	Pinwheel
25th Aug 1925	Arabian Sea	Moderate	Pinwheel
1943	Persian Gulf	Moderate	Pinwheel
11th Nov 1949	Strait of Hormuz	Strong	Pinwheel
7th Apr 1951	Gulf of Aden	Strong	Pinwheel
30th Nov 1951	Persian Gulf	Solid	Pinwheel
2nd Apr 1956	Persian Gulf	Moderate	Pinwheel

INDIA

Date	Place	Credibility	Details
5th Jun 1880	Malabar		Pinwheel
1887	Sikkim		Yeti Tracks
Dec 1914	Darjeeling		"Other" 1.3m hairy
1925	Sikkim		"Other" 1.3m hairy
Sep 1938	Lower Bengal		"Other" giant hairy
Feb 1942	Sikkim		2 "Others" giant hairy
1944	Kashmir		"Other" giant hairy
15th Mar 1951	New Delhi		UAP Cigar
15th Sep 1954	Bihar		UAP 4m saucer
2nd Nov 1954	Calcutta		UAP Bright disk
27th Nov 1954	Madras		UAO Boat flying
1954	Madras		UAP Disk
Late 1958	India		4"Others" 1.0m killed boy
16th May 1974	Bombay		UAP large saucer
24th May 1975	Arunachal Pradesh		2"Others" tall hairy caught
Summer 1977	Bombay		UAP gigantic disk
April 1982	Mysore		UAP 10m cigar 3 "Others" 1.7m gathering samples
Jan 1987	Kashmir		"Other" 1.3m hairy
20 Nov 1992	Calcutta		UAP Sphere
1st Mar 1997	Mumbai		"Other" giant bird
16th Aug 1996	Uttar Pradesh		"Other" like werewolf

EAST RUSSIA AND SEA OF OKHORSK

Date	Place	Credibility	Details
11th Jul 1908	Primorsky Krai		"Other" Tall, winged
22nd Nov 1908	Sea of Okhorsk	Moderate	Light underwater
Aug 1927	Khabarovsk Krai		UAP disk
Early 1943	Bering Sea		UAP running beside ship
Summer 1945	Adak Island, Alaska	Solid	UAP emerged
1951	Sea of Okhorsk	Moderate	Sonar detection of submerged UAP
Jan 1952	Shemya Island, Aleutian Islands	Solid	UAP entered sea
Sep 1956	Blagodatnoye Lake		UAP dish crashed
Jan 1957	Okhorsk		"Other" Tall woman
1957	Bering Sea	Solid	UAP out and in
1970	Kamchatka Lake		UAP emerged
1974	Kamchatka		UAP small sphere
14th Aug 1975	Kuril Islands		Abduction 2 weeks
1977	Sea of Okhorsk		UAP disk
24th Aug 1978	Khabarovsk Krai		Crashed UAP 6m mushroom
May 1982	Kamchatka		UAP 200m cigar
Summer 1982	Primorsky Krai		UAP disk
31th Oct 1987	Vladivostok		UAP huge disk Shot down
Spring 1989	Sea of Okhorst		UAP egg-shape into sea
20th Jun 1989	Primorsky Krai		"Other" Silvery grey
18th Jul 1989	Primorsky Krai		"Other" 2.0m
6th Sep 1989	Russkiy Island		UAP Rectangular
End 1989	Tikhaya Bay, Vladivostok		Triangular UAP
19th Jan 1990	Khanka Lake		UAP disk with windows
27th Mar 1990	Kuril Islands		UAP 1.3m sphere
31st Mar 1990	Primorsky Krai		UAP Oval
Apr 1990	Uyar		UAP
Summer 1990	Chukotka		UAP cigar & groups
Summer 1990	Primorsky Krai		"Other" 2.0m
Summer 1990	Vladivostok		"Other" Attempted Abduct
1990	Kamchatka		"Others" Tall spacesuits
1990	Timofeevka		"Others" 2.5m
14th Sep 1991	Vladivostok		UAP disk abduction
Nov 1991	Primorsky Krai		"Other" winged man
5th Sep 2006	Vladivostok		3 UAPs Spheres

BRITISH COLUMBIA COAST & SOUTHERN ALASKA

Date	Place	Credibility	Details
31st May 1831	Puget Sound WA		UAP luminous
2nd Jul 1893	Puget Sound WA		UAP 45m oval
Nov 1896	Vancouver BC		UAPs Luminous disks
1st Feb 1908	Puget Sound WA		UAP Brilliant Red
2nd Feb 1908	Puget Sound WA		UAP Red Cigar
14th Jan 1945	Oyster River BC		UAP silvery cylinder
5th May 1947	Seattle WA		UAP Silver object
18th Jun 1947	Eugene OR		2 UAPs shiny round
4th Jul 1947	Seattle WA		UAP Photo Dot
4th Jul 1947	Portland OR		3 UAPs disks
5th Jul 1947	Seattle WA		2 UAPs football
6th Aug 1947	Douglas OR		UAP 10m sphere
12th Nov 1947	Curry, OR		2 UAPs balls of fire
30th Oct 1948	Grays Harbour WA		20 UAPs eggs
24th May 1949	Curry OR		UAP circular
Sept 1952	Okanagan Lake AK	Solid	Surfaced UAP, taking off
16th Dec 1958	Hawk Inlet Juneau AK		UAP on surface
6th Mar 1960	Fraser River AK	Certain	Large UAP in & out
9th Mar 1960	Strait of Jan de Fuca WA		Flaming UAP into sea
Jun 1961	Seattle WA		UAP Cylinder
Jul 1963	North Portland OR		UAP 10m cylinder
12th Jan 1965	Tillamook OR		UAP Triangle Into sea
25th Mar 1966	Saceanich Inlet AK	rtain	2 UAP Triangles
3rd Apr 1966	Seattle WA		UAP large glowing
Apr 1966	Lincoln OR		"Others" Cyclops
Sep 1966	Lincoln OR		UAP ball of light
13thFeb 1967	Coos OR		UAP ball of light
Apr 1968	Vancouver Island BC		"Other" 1.5m
Spring 1968	West Vancouver BC		"Other" small glowing
Aug 1968	Vancouver BC		UAP 3m disk
10th Jun 1969	Portland OR		UAPs line of disks
Aug 1969	Douglas OR		UAP disk 1.0m
1st Jan 1970	Duncan BC		UAP 15m disk
May 1971	Portland OR		"Other" 3m long arms
Oct 1973	Duncan BC		UAP 25m disk
25th Jun 1974	Vashon Island WA		UAP 10m cylinder
End 1975	Lincoln OR		"Others" taking livestock
9th Mar 1977	Victoria BC		UAP Oval, light on top
5th Dec 1978	Ketchikan AK		UAP Large rectangle
10th Sep 1979	Douglas OR		UAP 10m sphere
25th Nov 1979	Grays Harbour WA		UAP Cigar

BRITISH COLUMBIA COAST & SOUTHERN ALASKA (Continued)

Date	Place	Credibility	Details
10th Dec 1979	Grays Harbour WA		UAP fireball crashed
7th Aug 1980	Tillamook OR		UAP rectangle
5th Oct 1981	Victoria BC		2 "Other" Men in Black
Oct 1981	Vancouver Island BC		UAP Photo high up
31st Aug 1987	Bangor Sub Base WA		UAP Disk with dome
15th Feb 1988	Rivers Inlet BC		UAP exploded
Spring 1995	Tillamook OR		"Others" Greys, orange eyes Mantis
15th Sep 1995	Wasilla Anchorage AK		2 UAPs orange lights
17th Nov 1996	East Vancouver BC		UAP Large black square
31st Mar 1997	Juneau AL		UAP coloured light
25th Apr 1997	Portland OR		"Other" Chupacabra
15th Mar 1998	Anchorage AL		2 UAPs conical
15th Sep 1998	Auke Bay Anchorage		UAP lights
7th Dec 1998	Anchorage AK		UAP lights
8th May 1999	Nikiski nr Anchorage		UAP light
25th May 2006	Ketchikan AK		30 oval UAPs
13th Sep 2010	Ketchikan AK		Spherical UAP
14th Sep 2010	Ketchikan AK		Multiple UAPs
28th Apr 2012	Ketchikan AK		UAP 3 lights
4th Sep 2012	Ketchikan AK		UAP above
8th Nov 2012	Ketchikan AK		UAP above
7th Sep 2013	Ketchikan AK		UAP plasma ball
5th Dec 2013	Ketchikan AK		UAP over airport
7th Aug 2014	Ketchikan AK		Oval UAP
19th Aug 2014	Ketchikan AK		Disk UAP
30th Dec 2014	Ketchikan AK		UAP above
11th Mar 2016	Ketchikan AK		Orange & Red UAPs

COASTAL CALIFORNIA

Date	Place	Credibility	Details
24th Feb 1904	USS Supply		3 UAPs Large disks
13th Sep 1930	Humbold		UAP fireball into sea
1935	Los Angeles		UAP & “Others”
Summer 1941	Ventura		Small silvery UAP
25th Feb 1942		The Battle of Los Angeles	
Nov 1943	Long Beach		UAP Glowing orange disk
July 1945	Los Angelrs		UAPs Darts of light.
21st Jun 1947	Maury I		5 UAP Debris
2nd Jul 1947	San Francisco		6 UAPs into sea. Footballs
3rd Jul 1947	San Diego		3 UAPs Disks
7th Jul 1947	Mendocino		UAP disk 2.0m
8th Jul 1947	Avalon		6 UAPs Flying Discs
9th Jul 1947	Santa Catalina Island		6 UAPs disks
Aug 1947	Los Angeles		UAP silver green sphere
20th Sep 1947	San Diego		UAP flaming, manoeuvred
May 1951	Solano		Multiple blasts
4th Jan 1952	San Diego		Several blasts
Jul 1952	Culver City		UAP Egg shaped
Mar 1953	Los Angeles		2 “Others” Tall, pricked ears
Nov 1953	Los Angeles		UAP Big silver ball
Nov 1953	Santa Monica		Squadron UAPs
29th Jan 1954	Santa Ana		UAP round with flat bottom
22nd Apr 1954	Saint Nicholas Island		UAP 2m cigar
11th Jul 1955	Santa Catalina Channel		UAP Saucer
22nd Jul 1955	Santa Maria	Weak	UAP Object Emerged
8th Aug 1955	Long Beach		UAP sphere in/out of sea
18th Jan 1956	Redondo		UAP Object Entered
9th Feb 1956	Redondo		UAP Sphere entered
20th Jul 1956	Los Angeles County		2 “Others” 2.2m blond
6th Nov 1957	Playa Del Rey		UAP Egg & 2 “Others”
14th Mar 1958	Sonora County		UAP 1m round
1958	Los Angeles		UAP Large Disk
19th May 1961	Long Beach		12 UAPs 2 explosions
28th Jul 1962	SE Catalina I	Strong	Object Surfaced
1964	Ventura	Strong	Object Emerged
5th Feb 1964	Humbold Bay		UAP under sea
1st May 1964	Santa Ana		UAP large domed
13th May 1964	Solano		UAP sphere
13th Sep 1964	Humbold		“Other” Big & Hairy
Nov 1964	Ventura		UAP from reservoir
2nd Dec 1965	San Pedro		UAP orange into sea
20th Aug 1966	Sonoma		UAP 15m high
Nov 1966	Palos Verdes		UAP triangle
28th Dec 1966	Santa Ana		UAP circular 6m

COASTAL CALIFORNIA (Continued)

Date	Place	Credibility	Details
1st Apr 1967	Sonoma		UAP domed
June 1967	Santa Barbara		7 "Others" 1m tall
Oct 1967	Catalina Strait	Solid	Object Entered
11 Oct 1967	Santa Ana		2 MIB
Spring 1968	San Diego		Explosion Red UAP
Oct 1968	Near Catalina I	Strong	UAP Emerged /Entered
4th Jan 1969	San Diego Mission Bay		UAP Egg shaped
24th Oct 1973	Los Angeles		UAP large round
Oct 1973	Santa Cruz		UAP hovered
12th Nov 1973	Los Angeles		UAP 10m ellipse
16th Nov 1973	San Diego		UAP 6m diameter 3m high
13th Dec 1973	San Diego		UAP 9m long 4m high
17th Jul 1974	Los Angeles County		UAP Cylinder
15th Oct 1974	San Diego		UAP round
Nov 1974	Los Angeles County		2 "Others" Tall &Hairy
Jun 1975	Santa Barbara		UAP 12m disk
Nov 1975	Los Angeles		UAP 0.6m sphere
Aug 1976	Bel Air		Several "Others" gorillas
16th Oct 1976	Ventura		UAP 3m oval
6th May 1977	San Diego		UAP Domed disk
25th July 1977	San Francisco		UAP conical
30th Aug 1977	Sonoma		UAP lights
4th Feb 1978	Orange County		UAP Oval
Summer 1978	Los Angeles County		UAP Oval 25-30m diameter
Nov 1978	San Diego		UAP Disk
15th Jun 1980	Santa Cruz		UAP 1.2m long cylinder
28th Nov 1981	Los Angeles County		UAP
27th Jan 1984	San Diego		5 UAPs lights in formation
Dec 1984	Santa Monica		UAP Ring of lights
9th Feb 1985	Los Angeles		2 UAPs orange spheres
May 1987	Santa Ana		UAP Silver ball
Sep 1987	San Luis Obispo Airport		UAP Intense light, radar
Oct 1989	Santa Rosa		UAP Light cluster
19th Sep 1991	Near Avalon		Multiple. NASA saw
14th Jun 1992	Catalina Strait	Certain	200+ Objects Emerged
30th Aug 1992	Malibu		Abduction
Feb 1993	Los Angeles		2 "Others" 1.3m
13th Mar 1995	Lake Takoe		UAP Triangle
Summer 1966	Beverley Hills		"Other" flying unassisted
8th Oct 1996	Ventura		"Other" Chupacubra
14th Nov 2004	Nr San Diego	Certain	Nimitz Incident
2005-06	San Francisco		UAP 2 triangles. 1 Disk
16th Feb 2008	Miramar		UAP Pulsing orb
22nd Jun 2009	Emerald Bay, Catalina		2 UAPs orb and submerged
14th Jan 2012	Orange County		UAP Orb
Jul 2019	USS Russell		UAP Triangles

WEST COAST OF MEXICO & CENTRAL AMERICA

Date	Place	Credibility	Details
30th Oct 1923	Costa Rica		Sea Monster long Neck
1941	Sonora Mexico		UAP Crash (before Roswell)
1944	Sonora Mexico		Dead "Other" short Large head
1945	Guatemala		3 "Others" 0.5m
1950	Honduras		"Other" Large Bat
1950	Panama		UAP Disk with dome
5th Mar 1950	Guerrero, Mexico		UAP Disk sighted by pilots
10th Mar 1950	Guerrero, Mexico		UAP Disk
May 1952	Honduras		UAP Disk "Other" Small Grey
9th Oct 1953	Jalisco Mexico		UAP Circular Small "Others"
Nov 1954	Costa Rica		UAP Globe
1958	Honduras		Round UAP "Other" <1.0m
30th Jul 1963	Sonora Mexico		3 UAPs huge white spheres
1965	Panama		UAP Large disk. Orange "Other"
30th Mar 1967	Guerrero Mexico		Cigar UAP crashed
1968	Honduras		Giant "Other" 5.0m
1973	Guatemala		2 "Others" Tall
1974	Jalisco Mexico		100m diameter UAP with fin
1974	Panama		"Other" <1.0m
22nd Dec 1975	Sonora Mexico		"Other" 1.3m
28th Oct 1976	Jalisco Mexico		200m diameter UAP
22nd Dec 1977	Guerrero Mexico		UAP 3x diameter of moon
17th Feb 1978	Jalisco Mexico		UAP Egg
18th Jul 1978	Jalisco Mexico		3 "Others" Amorphous
26th Nov 1978	Jalisco Mexico		2 UAPs luminous
1979	Costa Rica		UAP Circular
16th Jan 1979	Panama		UAP Oval
1st Jun 1983	Michoacan Mexico		UAP gigantic sphere
1983	Nicaragua		14 "Others" Tall
6th Feb 1984	Jalisco Mexico		5 UAPs Lights
1985	Jalisco Mexico		"Other" 1.2m normal head
22nd Mar 1986	Jalisco Mexico		"Other" 2.1m
25th Oct 1986	Costa Rica		UAP Underwater
22nd Nov 1986	Jalisco Mexico		"Other" golden figure
Dec 1986	Michoacan Mexico		"Other" tall thin blond
1991	Costa Rica		18 "Others" 1.7m
Apr 1993	Guerrero Mexico		UAP Disk
17th Aug 1995	Sonora Mexico		3 UAPs Orange "Others"
Mar 1966	Guatemala		"Other" Large Bat
1st Apr 1996	Chiapas Mexico		"Other" 0.5m killed sheep
1996	Honduras		UAP Box shaped
1997	Honduras		"Other" 1.3m Pointed hat
17th May 1997	Baja California Mexic		3 "Others" little men
24th Dec 1997	Costa Rica		12 UAPs disks
1999	Panama		"Other" 1.3m wrestled cattle

PERU

Date	Place	Credibility	Details
Oct 1907	Solimes River		"Other" long necked monster
16th Mar 1950	Lima		UAP glowing disk
19th Jul 1951	Iberia		UAP cigar
5th Aug 1952	Plane Lima to Huacho		3 UAPs saucer-shaped
May1954	Huamanga		"Other" 1.5m slender woman
30th Jan 1958	Arequipa		UAP saucer
2nd Feb 1958	Arequipa		UAP mushroom shaped
12th Apr 1960	Ancash		2 "Others" Tall
15th May 1960	Cordillera		2 "Others" Tall cured boy
1961	Catalina		UAP row of windows
2nd Aug 1965	Lima		"Other" Dwarf
20th Aug 1965	Cuzco		2 "Others" small from 1.5m disk
20th Aug 1965	Arequipa		"Other" 0.8m shrub entity
25th Aug 1965	Callap		UAP Saucer
3rd Sep 1965	Puno near Lake Titicaca		7 "Others" 1.0m Cyclops
12th Sep 1965	Huancavelica		2 "Others" 0.8m
17th Sep 1965	Huancavelica		UAPs 1 large 4 small disks
29th Sep 1965	Arequipa		"Other" 1.0m Cyclops
1st Nov 1965	Huanuco		UAP Ringed globe
5th Jul 1966	Lima		"Other" 1.5m
30th Dec 1966	Coast of Peru		UAP escorts aircraft
26th Mar 1967	Yungay		2 UAPs
2nd Jun 1968	Lima		6 UAPs soup plates
28th Nov 1968	Lima		"Other" short. Egg shaped head
3rd Feb 1972	Nazca		UAP cigar 15m, 2 Tall "Others"
11th Apr 1972	Cuzco, Lake Huaypo		UAP disk entered lake
Spring 1975	Lima		"Other" 2.2m tall
11th Nov 1975	Loreto		4 "Others" 0.8m UAP 4m oval
May 1977	Lima		0.5m "Other" UAP disk
18th Nov 1977	Arequipa, Yanyarina Sea		3 UAP disks entered sea
18th Nov 1977	Arequipa		3 giant "Others"
25th Dec 1978	San Martin		6 UAPs rhomboid
11th Feb 1980	Puno near Lake Titicaca		UAP disk
Apr 1981	El Collao		2 "Others" 1.2m Cyclops
1981	Paruro		"Others" 2.2m copper hair
1991	Lima		3 UAPs 1x100m 2x50m
1993	Puno		2 very tall "Others"
22nd Aug 1996	Ancash		3 UAPs Tall "Others" get water
Mar 1977	La Libertad		Huge UAP crash
20th Jul 1997	Corsica		UAP Cylinder 7m long
23 May 1999	Recuay		5 UAPs 30m
1999	Lima		"Other" small green head

CHILE

Date	Place	Credibility	Details
5th Nov 1883	Chile		UAP Large disk
13th Apr 1905	Magallanes		Long neck sea monster
Jan 1914	Malleco, Araucania		"Other" 2.0m angelic
Jul 1931	Los Lagos		UAP 50m on surface
Mar 1937	Antofagasta		UAP with lizard "Others"
1945	Antofagasta		UAP with small "Others"
10th Jul 1947	Santiago		UAPs oval
23rd Feb 1949	Chilean Antarctic		UAPs
23rd Feb 1950	Chilean Antarctic		UAPs
11th Mar 1950	Magallanes		UAP Sphere
1955	O'Higgins		Tiny "Other"
28th Oct 1958	Valparaiso		UAP
1958	Tarapaca		"Other" 3.0m thin
1958	Los Lagos		UAP Sphere small "Others"
9th Aug 1959	Magallanes		UAP 2m egg
Jan 1959	Los Lagos		3 "Others" 2.0m
Oct 1963	Malleco, Araucania		"Other" shape shifting woman
1963	Arica		UAP cigar
Nov 1964	Santiago		UAP disk very small "Others"
20th Jul 1965	Atacama		UAP mushroom-like
1964	Chiloe Isle, Largos		"Other" very tall in water
1964	Aracuo, Biobio		"Other" tall long blond hair
3rd Jul 1965	Chilean Antarctic		UAP lens shaped
4th Aug 1965	Santiago		UAP Oval landed
6th Sep 1965	Arica		UAP near aircraft
10th Sep 1965	Santiago		3 UAPs
11th Sep 1965	Villa Alemana, Valpariso		2 UAPs disks
21st May 1966	Colquinto		3 UAPs Spherical
1st Jun 1966	Bio-Bio		3 UAPs
Jun 1966	Los Morros, Santiago		UAP like nut
Jun 1966	Colquimto		2 "Others" 1.2m
4th Dec 1967	Santiago		"Other" Cat-like
May-Jun 1968	Antofagasta many times		UAP 3 "Ohers"
Jun 1968	Cordillera		Procession of "Others" in white
Jul 1968	Santiago		"Other" 0.7m
Jul 1968	Biobio		2 "Others" very tall
Aug 1968	Biobio		UAP disk
Sep 1968	Biobio		"Other" man with wings
27th Dec 1968	Santiago		UAP disk "Others" 0.8m
13th Feb 1969	Valdiva, Largos		UAP disk 3 "Others" 2.0m
10th Jun 1969	Santiago		UAP tube
22 May 1970	Chanaral, Atacana		UAP buzzed ship
24th Sep 1971	Antofagasta		Large Oval UAP
1972	Los Lagos		Large UAP in and out of water
21st Jun 1973	Santiago		"Other" bird man

CHILE (Continued)

Date	Place	Credibility	Details
11th Sep 1973	Araucania		Several UAPs
Jan 1974	Antofagasta		"Other" 2.0m
Summer 1975	O'Higgins		2 tall "Others" collecting fruit
3rd Aug 1975	Verdades ?		UAP bright sphere
1976	Los Lagos		UAP up and down in sea
25th Apr 1977	Arica		UAP and abduction
24th Aug 1978	Colquimto, at sea		UAP sphere. Web-fingered "Other"
29th Aug 1978	Huasco, Valpariso		3 "Others" 2.2m UAP Disk
26th Sep 1978	Santiago		5 "Others" short
16th Dec 1978	Antofagasta		Jet tries to intercept UAP
Nov 1979	Punta De Damas, Santiago		UAP Disk "Other" tall hairy
14th Jun 1980	Santiago		UAP sphere near aircraft
Jun 1980	Cordillera de los Andes ?		UAP 12m cigar. "Other" no neck
28th Apr 1983	BioBio		UAP over football match
1983 or 84	Antofagasta		2 Nordic "Others"
1983	Valparaiso		2 "Others" 2.5m
15th Sep 1985	Arica		2x 0.25m long strange creatures
Apr 1986	Atacama desert		80m UAP disc. Soldiers killed
1986	Valparaiso		2 "Others" 2.5m
Feb 1987	Elqui, Colquimto		"Other" large winged
31st Dec 1990	Santiago		Very tall "Other"
6th Nov 1993	Santiago		"Other" large winged
Oct 1994	Santiago		"Other" large winged
Summer 1996	Antofagasta		"Other" 3m thin long arms
16th Feb 1997	O'Higgins		"Other" 1.5m large head
31st Jan 1998	Colchagua, O'Higgins		UAP sphere blew up on ground
16th Jul 1998	Santiago		"Other" 0.5m hair on head
Aug 1998	Santiago		"Other" 1.9m slim
7th Oct 1998	Valle de Equi ?		UAP 15m hit mountain
23rd Jan 1999	Cautin, Araucania		UAP massive oval mothership
Jul 1999	Coquimbo		2 "Others" 2.0m Nordics
15th Sep 1999	Antofagasta		"Other" 1m Chupacabra
1999	Colquimto		"Others" small humanoids
2014	Tarapaca		UAP

NEW ZEALAND

Date	Place	Credibility	Details
1861	Lake Heron		Giant otter-like animal
Oct 1875	Waitahanui		3 "Others" 1m
1878	Waiki		"Other" big & hairy
May 1883	Masterton		"Other"
1883	Fortune Bay		"Other" Giant 20m Turtle
4th May 1888	Unknown		Oval UAP
14th Jul 1890	East Cape & Portland Light		"Other" 30m monster
May 1896	Wellington		"Other" White
23rd Jul – 9th Aug 1909	All over New Zealand		UAP Airship with gondola & 3 "Others"
1st Sep 1909	Kelso		UAP airship
1910	Coromandel		"Other" big & hairy
16th Jul 1920	Aukland		"Other" White
1919	Canterbury		UAP Ellipse
Aug 1944	Christchurch		UAP Saucer 3"Others" 1.1m
Feb 1952	Coromandel		"Other" big & hairy
6th Jan 1953	Auckland		UAP Cigar
8th Jan 1953	Mosgeil		UAP Triangle
2nd Jun 1953	Havelock		UAP 25m Cigar
19th Mar 1954	Nelson Bay		UAP Disk with dome
Nov 1954	Hamilton		"Other" 2.2m no neck green
6th Feb 1955	Greymouth		UAP Silvery Cigar
13th Nov 1955	Bluff		Cone & Cube "UAPs" surfaced
19th Jul 1956	Gisborne		Spherical UAP
Sep 1956	Putanuru		UAP 2m sphere
15th Jan 1957	Waiki & Balfour		UAP bright sphere or oblong
7th Jun 1958	Mount Egmont		UAP White oval
11th Apr 1959	Invercargill		UAP Sphere
3rd Nov 1959	Woodlands		UAP 12m long
3rd Jan 1960	Marlboroigh		UAP 10m sphere
16th Apr 1960	Napier		UAP vertical cylinder
Dec 1963	Coromandel		"Other" Apelike
12th Jan 1965	Kaipara Harbour		UAP submerged
Jan 1965	Waihoaka		UAP landed & departed
3rd Feb 1965	Canterbury		7m UAP
24th Mar 1965	Muriwai Beach		10m "Other" hairy body
13th Nov 1965	Rugged Islands , Southland	Solid	2 UAPs out and in
20th Jan 1968	Wellington		UAP Large silver disk
2nd Feb 1968	Auckland		UAP Domed disk Killed man
July 1968	Pacific Oan		"Other" 2m monster in net
10th Sep 1968	Hawkes Bay		UAP 0.75m sphere
29th Nov 1968	Manawatu-Wanganui		3 UAPs Lights in triangle

NEW ZEALAND (Continued)

Date	Place	Credibility	Details
18th Dec 1968	Wairakei		1.7m “Other” in helmet
22nd Feb 1969	Awanui		3 “Others” 1.7m
31st Aug 1969	Gisborne		UAP Disk
4th Sep 1969	Ngatea		Several UAP nests
4th Oct 1969	Katikati		2 UAP Orbs
8th Oct 1969	Dargaville		4 UAP nests, Cigar sighted
30th Oct 1969	Waipukurau		UAP circular metal
Oct 1969	Watford		UAP nest
Oct 1969	Rotorua		3 UAP nests
Oct 1969	Takapuna		2 UAP nests
7th May 1970	Richmond		UAP giant saucer, at school
9th Apr 1971	Canterbury		UAP
14th Apr 1971	Taupo		UAP disk with dome
1971	Lyttleton Fishing		“Other” hippo sized in nets
15th July 1972	Hamilton		UAP 15m Disk 3”Others”
1972	Coromandel Range		“Other” tall & Hairy
8th Jan 1975	Auckland		UAP domed Disk
12th July 1975	Dunedin		“Other” 1.75m
5th Feb 1977	Auckland		2 “Others” human shape & size
April 1977	Gisborne		UAP domed disk “Others”
April 1977	Nr Christchurch		Fishers net 10m dead monster
July 1977	Kaiapot		“Other” big cat
7th Dec 1977	Gisborne		UAP disk
2nd Jan 1978	Auckland		UAP 2 “Others” 2.2m
30th May 1978	Auckland		UAP small cigar
16th Jul 1978	Auckland		UAP rectangle & 2
16th Dec 1978	Marlborough		UAP inverted wheatsheaf
1980	Gisborne		4 “Others” humanlike
11th Jul 1981	Dunedin		UAP football
12th Jul 1981	Canterbury		UAP half-oval
2nd Aug 1982	Canterbury		“Other” silver 2m
1st Jun 1994	Christchurch		UAP 4 lights
10th Jun 1995	Northland		UAP diamond
7th Feb 1998	Kapiti Island		UAP 4 green lights cylinder
26th Nov 1999	Auckland		UAP cigar

AUSTRALIA, PAPUA & NEW GUINEA

Date	Place	Credibility	Details
1821	Lake Bathurst		"Others" large Amphibians
1822	Sydney Cove		"Other" tall hairy Yowie
1828	Mount Wingen		UAP cigar
1844	New South Wales		"Other" tall hairy Yowie
1860	South Australia		UAP 18m orb
1872	Lake Burrumbert		"Other" Bunyip
1885	Victoria		"Other" Tiger-like
1885	Blue Mountains		UAP Disk 2 "Others"
1896	Bass Strait		UAP airship
July 1909	Blue Mountains		UAP cigar
15th Aug 1909	Perth		UAP airship
1910	Karumba		UAP disk abduction
Nov 1912	Currockbilly Mountains		"Other" 1.7m red hair, fangs
21st Jul 1913	Bathurst		"Other" 1.0m dwarf
1918	Watagan Mountains		"Other" 4m lizards
Jan 1924	Melbourne		UAP 6 Disks
1927	New South Wales		"Other" 3x long nose UAP
1930	Western Australia		"Other"0.5m pink flesh
18th Jun 1931	Port Augusta Channel	Solid	UAP on surface
1933	Western Australia		UAP egg
Feb 1934	Queensland		"Other"1.7m long arms
1938	Queensland		"Other" 1.8m Fair, no neck
Jan 1942	Tasman Peninsular		UAP Disk with dome
Feb 1944	Bass Strait		UAP Disk with dome
Summer 1944	Tasman Peninsular		UAP 30m long
6th Feb 1947	Port Augusta		3 UAPs egg shaped
Feb 1948	Berridale		UAP 3 small green "Others"
1948	Eggletons Plain		UAP 70m cigar
Oct 1949	Townsville		UAP 35m cigar
May 1950	Kings Tableland		UAP14m egg
3rd May 1952	Melbourne		UAP silver cigar
18th Oct 1952	South Australia		UAP cigar 50m
1953	New South Wales		Several saucer nests
8th Jan 1953	Queensland		UAP 1m light
23rd May 1953	Melbourne		UAP 30m cigar

AUSTRALIA, PAPUA & NEW GUINEA Continued)

Date	Place	Credibility	Details
23rd Aug 1953	Port Moresby, New Guinea		UAP Dart shape
1953	New South Wales		UAP 9m disk
5th Feb 1954	Western Australia		UAP Disk
13th Mar 1954	South Australia		UAPs many multicolour
7th Apr 1954	Brisbane		UAP cigar
9th Apr 1954	South Australia		UAP long silver cigar
18th Apr 1954	Western Australia		UAP Disk with portholes
1st May 1954	Melbourne		UAP giant football
5th May 1954	Melbourne		UAP 30' cylinder
30th May 1954	Melbourne		UAP Oval size of a tram
4th Jun 1954	Melbourne		UAP green sphere
9th Jun 1954	Victoria		UAP 10m cylinder
5th Jul 1954	Melbourne		UAP cigar size of a car
31st Aug 1954	Over New South Wales		2 UAPs radar & pilot
9th Sep 1954	Victoria		UAP Disk
1954	Bankstown airport		UAP disk with portholes
Spring 1955	Balook		3 "Others" 23 cm tall
2nd Aug 1955	Eucla		"Other" Frog back
Aug 1956	Mount Pleasant		3 "Others" Cylinder UAP
1957	Bayswater		"Other" 1.4m
13th Jan 1958	Sydney		UAP huge dome
1958	Townsville		4 "Others" 2m fair
Jan 1959	Gulf Saint Vinnt	Solid	2 UAPs emerged
13th Mar 1959	Purnong Landing		UAP domed
26th Jun 1959	Papua & New Guinea		UAP over Mission
17th Mar 1960	New South Wales		UAP large candlestick
4th Oct 1960	Tasmania		5 UAPs, one large
6th Aug 1961	Western Australia		12 UAPs in pairs
20th Oct 1961	Sidney		UAP 60m long
1962	Bass Strait	Solid	UAP on sonar
16th Feb 1963	Moe, New South Wales		UAP Thick Disk
Spring 1963	Plympton		UAP disk 2 Tall "Others"
23rd Jan 1964	Gulf of Carpentaria	Certain	Pinwheel
16th Sep 1964	Tasmania		UAP attacks car
Jan 1965	Brisbane		100m Wide UAP
24th May 1965	Nr Nebo, Queensland		UAP circular, landed
9th Jan 1966	Tully, Queensland		UAP. Flying saucer nests
6th Apr 1966	Westall, Melbourne		UAP near school
24th Aug 1967	Woodonga, Victoria		UAP chased M/C
Dec 1967	Queensland		3 small "Others" in silver
3rd April 1968	Narrabeen Lakes		1.2 "Other" Elephant legs
10th May 1968	Tasman Sea	Strong	UAP emerged
1968	Brisbane		UAP Black Triangle
14th Jan 1969	Childers		UAP 3 giant "Others"
1969	Cairns		"Other" 2 tall hairy& child
1969	Tasmania		UAP Oval Tall, thin "Other"
8th Dec 1970	Western Australia		UAP silver disk, portholes

AUSTRALIA, PAPUA & NEW GUINEA Continued)

Date	Place	Credibility	Details
3rd May 1971	New South Wales		"UAP" entered sea
May 1972	New South Wales		3 "Others" 1.0m Tall
23rd Jun 1972	New South Wales		UAP cigar
3rd Jul 1972	Frankston Victoria		Large UAP saur
30th Apr 1973	New South Wales		UAP egg
Nov 1973	New South Wales		UAP 5m egg
6th Mar 1975	Tasmania		UAP Hexagonal
22nd Mar 1975	Nr Nebo, Queensland		UAP disk landed
19th Oct 1975	New South Wales		5 "Others" in UAP cigar
12th May 1976	New South Wales		UAP Disk crashed in sea
17th Oct 1977	Western Australia		"Other" 1.6m slender
Dec 1977	New South Wales		"Other"1.5m no neck
21st Oct 1978	Bass Strait		Large UAP hit aircraft
Oct 1978	Queensland		"Other" 1.0m black hairy
21st Feb 1979	Queensland		UAP cigar
20th Oct 1979	Victoria		UAP in field beside road
29th Sep 1980	Gippsland, Victoria		UAP drained water tank
2nd May 1982	Queensland		UAP Disk. Many Nordics
Jan 1983	Western Australia		"Other" 1.0m
1983	New South Wales		"Other" Avian 2.0m
Feb 1984	Victoria		"Other"1.5m no neck
Jul 1984	New South Wales		UAP 100m Ovoid
26th Dec 1984	New South Wales		4UAPS 1 cigar, 3 Disk
1985	New South Wales		"Other" 1.0m bulky
Jan 1987	Tasmania		"Other" 1.8m no neck
20th Jan 1988	Nullabor, W Australia		UAP attacks car
Aug 1991	South Australia		UAP Cone 1.3m "Other"
14th Oct 1991	New South Wales		UAP emerged
22nd Dec 1995	Western Australia		Abduction 2300miles 10min
Oct 1996	South Australia		UAP disk 70m
Dec 1997	Western Australia		"Other" Giant Flying

MALAYSIA & INDONESIA

Date	Place	Credibility	Details
16th Jul 1864	Gulf of Thailand	Moderate	2 Pinwheels
Oct 1891	South China Sea	Weak	Searchlight rays below
14th Mar 1907	Strait of Malacca	Moderate	Pinwheel
10th Jun 1909	Strait of Malacca	Solid	Pinwheel
12th Aug 1910	South China Sea	Moderate	Pinwheel
24th Aug 1911	Gulf of Thailand	Solid	Light out/in water
Jun 1930	Gulf of Martaban	Moderate	Pinwheel
1950	Malaysia		3 UAPs disks
23rd Oct 1952	Strait of Malacca		3 UAPs orange spheres
Sep 1958	Borneo		"Other" 6m
1959	Penang		"Other" Tiny man
1960	Perak		"Other" tiny family
1962	Penang		4 "Others" 1.7m
30th May 1962	Gulf of Siam	Strong	Pinwheel
1964	Pahang		2 "Others" tiny
1965	Perak		3-4 "Others" tiny
Aug 1966	Johor		"Other" 8m
1966	Selangor		2 "Others" 0.3m
2nd Jul 1969	Johor		5 "Others" tiny
19th Aug 1970	Penang		5 "Others" 7.5 cms
20th Aug 1970	Penang		2 "Others" 7.5 cms
24th Aug 1970	Selangor		5 "Others" 7.5 cms
1973	Perak		UAP tiny Several "Other" tiny
9th Oct 1974	Pahang		3 tiny "Others"
1976	Pahang		"Other" tiny
1978	Pahang		Several "Others" tiny
Mid 1979	Pahang		3 tiny "Others"
18th Jun 1980	Perak		3 tiny "Others"
24th Jan 1982	Perak		UAP Saturn shaped
21st Aug 1982	Sarawak		"Others" large heads 0.8m
Aug 1984	Selangor		4 "Others" large heads 0.8m
11th Oct 1985	Terengganu		Several "Others" 0.1m
27th Jul 1986	Terengganu		UAP 3m sphere. 1m "Other"
13th May 1991	Terengganu		"Hundreds" of tiny "Others"
Sep 1995	Selangor		UAP football pitch size
Dec 1996	Selangor		3 "Others" large heads
May 1999	Sarawak		Several tiny greenish "Others"

JAPAN & THE DRAGON'S TRIANGLE

Date	Place	Credibility	Detail
1942-45	Japan		UAPs Foo fighters
April 1945	Okinawa		UAP 12m cigar
15th Oct 1948	Japan		UAP chased by US fighter
21st Nov 1949	Akita		UAP rectangular
27th Aug 1950	Hachinohe		UAP flat 20m long
19th May 1950	Honshu		UAP Triangular
Dec 1950	Yellow Sea	Strong	2 UAPs entered
15th Feb 1951	Sea of Japan		5 UAPs on radar
9th Mar 1951	Japan-Korean Airspa		UAP radar
4th Mar 1952	Ashiya AFB		UAO disk
22nd Apr 1952	Okinawa		5 UAPs lights
30th May 1952	Oshima Island		UAP
14th Jul 1952	Okinawa		UAP sphere
5th Aug 1952	Haneda AFB		UAP on radar, sighted
14th Mar 1953	Hiroshima		90-100 UAPs
29th Dec 1953	Japan		UAP radar pilot sighting
12th Aug 1954	27°N 128°E		100 ft UAP
24th Mar 1955	Ryukyu Islands		UAP Disk with windows
15th Jan 1956	Sea of Japan	Strong	UAP entered
7th Sep 1956	Chiba		UAP disk – angel hair
9th Nov 1956	Yokohama		Several UAPs, thousands
17th Dec 1956	Itazuke		UAP on radar too fast
17th Jan 1957	Yokohama		UAP photo
21st Feb 1957	Yokohama		UAP fleet
Mar 1957	Le Shima Island		UAPJet crashed into saucer
13th Mar 1957	Tokyo		UAP cigar
19th Apr 1957	Pacific, near Japan	Solid	2 UAPs entered
10th Jun 1957	Yokohama		UAP cigar
20th Aug 1957	Fujisawa		UAP cigar, photo
10th Nov 1957	Lake Imba-numa		UAP cigar
26th Jan 1958	Shimada		UAP lands "Others"
2nd Feb 1958	Hokkaido		UAP egg shaped
17th Feb 1958	Yokohama		UAP cigar photo
26th Jan 1959	Hokkaido		2 UAPs
8th Apr 1961	34°N 132°E		Cigar UAP
21st Mar 1965	Osaka		UAP approached aircraft
11th Jan 1966	Western Pacific	Certain	Large UAP entered
15th Feb 1966	Osaka		UAP saucer
Sep 1966	Western Pacific	Solid	Light underwater
5th Jul 1968	Philippines	Solid	Conical UAP entered
1970	Mount Hiba		"Other" 1.3m hairy
Oct 1972	Hiroshima		"Other" 1.5m hairy
9th Jun 1974	Japan		UAP disk hit jet

JAPAN & THE DRAGON'S TRIANGLE

Date	Place	Credibility	Detail
23rd Feb 1975	Kofu		UAP landed 1.3m with fangs
15th Mar 2009	Japan		Several UAPs
16th Sep 2014	Japan		Triangular UAP

COASTAL RUSSIA

Date	Place	Credibility	Details
9th Aug 1845	Bornholm, Baltic Sea		Very Large Flame
16th Sep 1897	Arkhangelsk		UAP shining disk
30th Jun 1908	Tunguska, Krasnoyarsk		Massive explosion
Jun 1908	Krasnoyarsk		"Other" 1.2m dwarf
1920	Arkhangelsk		"Other" Hairy giant 6 toes
1947	Baltic Sea		Several Spherical UAPs
1950s	Black Sea		UAP emerged
16th Jul 1951	Murmansk		UAP disk 20m
Sept 1956	Sea of Japan		Enormous UAP crashed
Sep 1957	Black Sea		Spherical UAP
Jun 1958	Aral Sea		Disk UAP
1960	Barents Sea	Weak	UAP emerged
1961	Krasnoyarsk sky		UAP disk Passengers disappeared
Aug 1965	Black Sea		UAP Disk
Sep 1965	Kola Peninsular		2 UAPs one shot down other
1966	Black Sea	Weak	UAPs above navy
1967	Bay of Viborg		UAP Disk
Nov 1967	Baltic, nr Latvia		Gigantic UAP hemisphere
Aug 1970	Black Sea, Sochi		12 UAPs rectangles
1972	Krasnoyarsk		UAP sphere
7th Oct 1977	Barents Sea		9 UAPs circle ship
1978	Laptev Sea		UAP domed
28th Aug 1978	Kola Peninsular		3 "Others" tall
1979	Zapadnaya Sub Base		Disk UAPs overfly base
1979	Zapadnaya Sub Base		Disk UAP flies over dived sub
May 1979	Black Sea		Gigantic UAP Disk, exit
12th Aug 1979	Black Sea		100m UAP Sphere
Oct 27th 1979	Khatanga		Huge UAP & 3 spheres
17th Dec 1979	Khatanga		Multiple Sphere UAPs
End 1982	Black Sea Balaklava		UAP entered
Summer 1983	Black Sea, Bathyscape		Tall thin "Other" big eyes
Nov 1983	Kola Bay		3 UAP spheres
1984	Baltic		3m "Others" down 400m
Oct 1984	Laptev Sea		Radar Sphere UAP sighted
23rd Oct 1985	Arctic Oan		4 UAPs
July 1986	Arkhangelsk		UAP disk
11th Aug 1987	Black Sea		Triangular UAP
Sep 1988	Murmansk		"Other" Tall hairy
1989	Kil'din I, Olenegorsk		3 UAPs
Jul 1989	Black Sea		Cigar UAP & small square
2nd Aug 1989	Primorskiy Krai		Sphere UAP
2nd Nov 1989	Arkhangelski		UAP disk 45m
6th Nov 1989	Krasnoyarsk		"Other" 2.2m no neck
1989	Murmansk		"Other"
21st Mar 1990	Krasnoyarsk		2 huge UAPs

COASTAL RUSSIA (Continued)

Date	Place	Credibility	Details
Apr 1990	Kradnoyarsk		Several UAPs
Spring 1990	Franz-Joseph Land		UAP
Summer 1990	Murmansk		UAP disk 4 "Others"
1991	Arkhangelski		UAP sphere with ring
Nov 1992	Arkhangelski		UAP disk
Winter 1993	Odessa, Ukraine		UAP on surface
Summer 1995	Murmansk		"Other" Tall winged
13th Aug 1995	Karsk Sea		Submerged UAP
1996	Kirenniy3ni Cape		UAPs Spheres & Cylinders
1997	Black Sea		Bell UAP "borrowed" torpedo
Summer 1997	Cape Aiya, Crimea		2 very tall "Others" swimming
23rd Aug 1999	White Sea		"Other" pillar of light grew 200m
1999	Kola Peninsular		"Other" small

HAWAII

Date	Place	Credibility	Details
12th Aug 1825	At sea		UAP Red sphere in and out
18th Jun 1944	Oahu		UAP sphere 7 "Others 1.5m
18th Oct 1948	Oahu		UAP ellipse 5m
17th Jan 1950	Kauai		UAP disk
Apr 1950	Kauai		UAP 6 red lights
14th Mar 1952	From aeroplane		2 UAP disks
6th Mar 1953	Naval Air Station		75 UAPs
7th Jan 1956	Honolulu		UAP disk 35m
12th Mar 1963	Oahu		UAP Red sphere
6th Jun 1970	Maui		Many UAPs
15th Jul 1972	Kauai Naval Base		UAP forced down by Tomcats
1972	Maui		"Others" Nordics & Helmeted
1973	Wahiawa		"Others" Bigfoot, Giant woman
1974	Maui		Woman abducted permanently
1977	Maui		UAP attempted abduction of 4
1977	Maui		UAP 350m saucer
Jul 1980	Oahu		"Other" Dwarf
Aug 1980	Oahu		UAP mother ship & smaller
4th Sep 1994	Oahu		UAP 10x size of Aircraft Carrier
1996	Oahu		"Other" Menehune 1.0m

NOVA SCOTIA, LABRADOR & NEWFOUNDLAND

Date	Place	Credibility	Details
8th April 1813	Nova Scotia	Solid	UAP on surface
15thSept 1850	Wellington Channel, Canada	Strong	UAP above surface
12th Nov 1887	Cape Ra, Newfoundland	Solid	UAP emerged
Aug 1914	Georgian Bay, Canada	Moderate	UAP emerged
1932	OveeGreenland		UAP hexagonal disk
9th Jul 1947	Grand Falls, Newfoundland		4 UAPs round
10th Jul 1947	Harmon Field NFL		UAP disk
14th Aug 1947	Harmon Field NFL		2 UAP crescents
28th Oct 1948	Goose Bay Labrador		2 small UAPs on radar
21st Apr 1950	Cape Cod		UAP overtook jet
30th Jul 1950	St Johns NFL		UAP 3m long
4th Aug 1950	S of Nova Scotia	Certain	UAP above surface
10th Feb 1951	Nr Newfoundland		UAP emerged
Early 1952	Goose Bay Labrador		UAP sphere
April 1952	Labrador Sea	Solid	Sounds underwater
18th Aug 1952	Gander		UAP near miss
29th Aug 1952	Thule Greenland		3 UAPs silvery disks
23rd Oct 1952	Seal I. Nova Scotia		UAP light in water
24th July 1953	Greenland		UAP hits balloon
12th Oct 1953	Cape Cod		UAP destroyed jet
10th Nov 1953	Greenland		Large UAP
28th July 1954	Cape Cod	Certain	UAP emerged
24th Sep 1960	North of Labrador		UAP Cylinder entered
6th Oct 1961	Newfoundland Nuclear Site		UAP 4 minutes
29th Nov 1965	Cooper Creek, Nova Scotia		UAP dome with portholes
21st Nov 1966	Prince Edwards Island		UAP shiny disk
Summer 1967	New Brunswick		"Other" Knee-high
15th Sep 1967	R Cornwallis, Nova Scotia		UAP disk 4.5m
4th Oct 1967	Shag Harbour, Nova Scotia	Certain	UAP crash landed
11th Mar 1968	Marshalltown, Nova Scotia		UAP Oval
4th May 1968	Clark's Harbour, Nova Scotia	Solid	2 UAPs repairing
15th Sep 1968	Halifax, Nova Scotia		UAP domed 4.5m
25th Nov 1970	Shag Harbour, Nova Scotia		UAP lights
Sep 1973	Oak Island, Nova Scotia		"Others" 1 small, 1 taller
28th Jun 1975	Halifax, Nova Scotia		UAP 7m disk
3rd Aug 1976	Halifax, Nova Scotia		UAP 10m Oval
26th Oct 1978	Labrador		UAP Circular 2 hours
21st Oct 1983	Godhavn, Greenland		UAP oval dome
Summer 1987	New Brunswick		"Other"0.25m swimming
25th Jan 2010	Harbour Mile		Multiple UAPs
4th Jun 2014	Prince Edward I		Lights 22 mins

BERMUDA, FLORIDA AND THE SOUTHERN EAST COAST OF USA

Date	Place	Credibility	Details
27th Aug 1885	Bermuda		UAP triangular
29th Aug 1890	Nr N Carolina	Solid	UAP above
Feb 1894	W Coast of Florida	Moderate	UAP emerged
9th Feb 1913	Hamilton		UAP train heading south
27th Jun 1944	St Mary's Maryland		"Other" large winged
18th Jul 1945	Miami Naval Air		Plane missing
5th Dec 1945	Miami Naval Air		Flight 19 missing
5th Dec 1945	Miami Naval Air		Flying Boat missing
Jul 1946	Sussex Delaware		UAP disk in and out of sea
18th Nov 1948	Prince George Maryland		UAP highly manoeuvrable
28th Dec 1948	Miami Airport		DC3 disappeared
1950	Duval		UAP 10m 3 "Others" greys
4th Aug 1950	39°N 72°W		UAP circular
11th Jun 1952	33°N 75°W		UAP small square
16th Jul 1952	Wicomico, Maryland		UAP Bright light
Sept 1954	Nr Georgia	Solid	UAP emerged
22 Aug 1957	Miami Naval Air		UAP 15m bell
July 1959	North Carolina		Loud explosion in sky
24th Aug 1959	28°N 85°W		2 white lights
30th Sep 1959	36°N 70°W		Stationary light
1959	USS FDR Florida	Certain	UAP oblong
6th Aug 1960	29°N 79°W		White/green light
11th Jul 1961	30°N 76°W		Diamond UAP
15th Jan 1962	38°N 74°W		9 UAPs
29th Jan 1962	32°N 67°W		5 UAPs
12th Aug 1963	Nr Miami	Solid	UAP emerged
23rd Jun 1964	Palm Beach		UAP 2m
Summer 1965	Nr Miami		UAP disk
Oct 1965	Worcester, Maryland		4/5 "Others" egg heads
6th May 1966	Dade County		UAP 30m
5th Jun 1966	Brevard County		2 "Others" UAP submerged
1st Aug 1966	Prince George Maryland		UAP fast light, on radar too
Aug 1966	Sussex, Delaware		UAP landed
9th Feb 1967	Newcastle, Delaware		UAP Disk with cupola
19th Feb 1967	Broward County		UAP oval
3rd Mar 1967	Broward County		2 UAPs cones
Mar 1967	USS Annapolis		UAP Cigar
20th Jul 1967	Brevard County		UAP domed disk 20m
20th Mar 1968	Nr S Carolina	Solid	UAP emerged
2nd May 1972	Wicomico, Maryland		UAP disk. "Others" 1.2m
8th Sep 1973	Chatham, Georgia		UAP light
18th Oct 1973	Savannah Georgia		"Other" small
Jul 1974	West Palm Beach		UAP disk
22nd Aug 1978	Southampton		UAP 10m disk
21st Feb 1979	Nr Miami		UAP top shaped

BERMUDA, FLORIDA AND THE SOUTHERN EAST COAST OF USA (Continued)

Date	Place	Credibility	Details
28th Sep 1980	Over Palm Beach		2 UAPs disk buzzed plane
30th Jan 1984	Oil Rig near Miami		UAP 70m domed disk
Aug 1988	Miami		"Other" tall
May 1993	Dade County		UAP 30m Green "Other"
16th Mar 1995	Palm Beach		Several UAP disks
Dec 1995	West Dade County		3"Others" tall blond
6th May 1999	Brevard County		UAP 100m oblong

GULF OF MEXICO & COASTLINE

Date	Place	Credibility	Details
14th Jul 1841	At sea		"Other" Long neck monster
13th May 1872	At sea		"Other" Long neck monster
21st Jun 1908	At sea		"Other" Long neck monster
6th Sep 1946	New Orleans LA		UAP Red ball
6th Jul 1947	Galveston TX		7 UAPs disks
27th Jan 1949	Manatee, Florida		UAP 40m cigar
1st Apr 1952	320km s off Louisiana		UAP fell into sea
6th Sep 1955	Levy, Florida		6 UAPs
1957	Deep under oil rig		"Others" green frog-like
13th Aug 1959	Brazoria TX		UAP very bright
1961	New Orleans LA		"Others" 1.2m
July 1962	Neus TX		UAP submerged
3rd Sep 1965	Brazoria TX		UAP 60m cigar
Sep 1965	Veracruz, Mexico		UAP 7m
6th Feb 1966	Jefferson TX		UAP 4m Tadpole
26th Jul 1966	Yucatan, Mexico		UAP large disk
6th Apr 1967	Okaloosa, Florida		UAPs over school
May 1967	Hernando, Florida		"Other" Tall hairy
June 1967	Veracruz, Mexico		"Other" Tall hairy
27th Mar 1968	At sea		UAP fiery ball manoeuvred
Aug 1968	Pinellas, Florida		UAP 2m diameter globe
21st Aug 1968	Hernando, Florida		UAP Large globe
28th Oct 1969	Mobile Alabama		UAP 15m disk
Jun 1970	Veracruz, Mexico		"Others" 1.0m abducted
March 1973	Veracruz, Mexico		"Others" 1.0m abducted
11th Oct 1973	Pascagoula MS		UAP egg 3"Others" 1.5m Abdu
16th Oct 1973	Gulport MS		UAP
6th Nov 1973	Pascagoula MS		3m Submerged UAP
Jan 1974	Harrison MS		UAP Large Ovoid
15th Dec 1974	Jefferson TX		UAP disk
11th Nov 1975	New Orleans LA		Large disk UAP. Tall "Others"
30th Oct 1978	Aransas TX		UAP Triangle
1978	Veracruz, Mexico		"Other" Large Bat
Autumn 1979	Veracruz, Mexico		"Other" 3m tall thin
Apr 1986	Cameron TX		UAP disk
Spring 1988	Pensacola, Florida		UAP crash dead "Others" greys
6th Jan 1989	Pensacola, Florida		UAP Orange ball
9th Apr 1989	Pinellas, Florida		7 UAPs disks 2-3m
Apr 1989	Veracruz, Mexico		UAP grey sphere crash
11th Aug 1989	Brazoria TX		"Others" lizards Abduction
Apr 1990	Santa Rosa, Florida		6 "Others" Abduction

GULF OF MEXICO & COASTLINE (Continued)

Date	Place	Credibility	Details
19th Jan 1995	Quintana Roo, Mex		3UAPs fiery balls manoeuvred
27th Oct 1995	Veracruz, Mexico		2 UAPs collision
Nov 1995	Veracruz, Mexico		UAP disk with dome
Feb 1996	Tamaulipas, Mexico		"Other" short red hair claws
3rd May 1996	Sinaloa, Mexico		"Other" Giant bat
20th Sep 1996	Quintana Roo		3 UAPs orange lights
Dec 1996	Yucatan, Mexico		"Other" very tall and thin
August 1997	Hernando, Florida		2 "Others"
17th Nov 1997	Pinellas Florida Sea		2 UAP amber lights
13th May 1998	Okaloosa, Florida		UAP Triangle

PUERTO RICO, BERMUDA TRIANGLE & SOUTH CARIBBEAN

Date	Place	Credibility	Details
Jul 1942	Toa Alta		UAP 5m disk
Early 1946	Barrio Utuado		"Other" 1m tall
14th May 1952	Mayaguez		2 UAP orange spheres
6th Oct 1952	Cabo Rojo, Puerto Rico	Solid	UAP entered
31st Dec 1952	Ramsey AFB		UAP Orange sphere
1956 -1960	Laguna Cartagena		Multiple UAPs
18th Jan 1956	Nr Venezuela	Solid	Large UAP landed
20th Feb 1956	Vieques Island		UAP 10m Disk
13th Mar 1963	Nr Puerto Rico	Moderate	UAP on sonar
11th Apr 1963	Puerto Rico Trench	Solid	UAP under
Summer 1963	Between Puerto Rico & Florida	Sold	UAP on sonar
Spring 1967	El Yunque		"Other" Tall thin
1968	Bermuda triangle	Solid	UAP on surface
June 1968	Nr Puerto Rico	Moderate	2 UAPs in|out sea
July 1968	El Yunque		"Other" Tall thin
31st Dec 1968	Vega Baja		"Other" 1.7m
Summer 1971	El Yunque		"Other" Tall
28th Oct 1972	Rio Pedras		"UAP" Large oval
Nov 1972	Culebra Island		"Other" 1.3m reptilian
20th Oct 1973	El Yunque		3 "Other" Canine
2nd Nov 1973	Pina		UAP Oval
3rd Jan 1974	Bayamon		UAP Disk
Dec1974	El Yunque		1m "Other" trapped
Mar 1975	Moca		"Other" Large Bird
12th Mar 1975	Moca		UAP big as a house
18th Apr 1975	Pon		"Other" Dwarf
1975	Caguas		UAP Egg
Jun 1977	Quebada Granda		"Other" big head eyes
31st Aug 1977	Mediana Alta		UAP windows, 4 small
17th Jul 1978	Camuy		"O" 2 Greys & 1 Tall
Apr 1979	Cabo Rojo		"Others" 3 Small Greys
Dec 1979	El Yunque		"Others" Several Greys
Summer 1979	San Juan		"Others" Large Bird
1982	El Yunque		2 "Others" 1.0 m
20th Nov 1983	Sierra del Yunque		"O" family of 3 tall hairy
Nov 1984	El Yunque		"Other" Nordic
Sunmer 1985	El Yunque		"Other" 2 giant birds
9th Jul 1986	El Yunque		7 "Others" Nordic
31st May 1987	Laguna Cartagena		UAP Huge Saucer
Late 1987	Laguna Cartagena		5 "Others" Small greys
28th Dec 1988	Laguna Cartagena		UAP Large Triangle
Jul 1989	El Yunque		Large hole in ground
24th Sep 1971	Laguna Cartagena		UAP Disk "O" 7 small G

PUERTO RICO, BERMUDA TRIANGLE & SOUTH CARIBBEAN (Continued)

Date	Place	Credibility	Details
Oct 1992	El Yunque		"Other" 1.1 silky hair
1992	Maracal		"Other" 1.1 silky hair
May 1993	Vieques Islanda		UAP Manta Ray 15m
1994	San Juan		"Other" Large bird
7th Apr 1995	Orocovis		"Other" Grey with crest
11th May 1995	San Juan		"Other" 1.0m bird
Aug 1995	Dorado		"Other" Chupacabra
Dec 1995	Caja de Muertos		Large UAP
Feb 1996	Dorado		"Other" Chupacabra
2nd Apr 1996	Adjunta		3 UAPS large disks
4th Aug 1967	Nr Venezuela	Strong	UAP disk emerged
27th Aug 1967	Nr Venezuela	Solid	3 UAP emerged
12th May 1998	US Virgin Islands		6 UAPs orbs
25th Jan 2006	St Maarten		2 saucer UAPs
23th Apr 2013	Rafael Hernandes Airport		UAP in/out
8th Jul 2014	Culebra Island		UAP
20th Dec 2014	Cayeu		2 UAPs
27th Apr 2016	Northwest Coast		2 UAPs
1st Oct 2016	Puerto Rico		Saucer UAP

CUBA & WEST CARRIBEAN

Date	Place	Credibility	Details
Feb 1947	Havana		UAP Airship
1947	Kingston Jamaica		2 UAPs silver disks
9th Mar 1950	Santiago		UAP Disk
7th Apr 1952	18°N 65°W		Large circular UAP
Summer 1957	Bacuranao	Moderate	UAP emerged
17th Sep 1958	16°N 76°W		Large UAP
Autumn 1958	Guantanamo Bay		UAP Large disk
9th Oct 1958	Nr Cuba	Strong	Large UAP Cylinder
16th Sep 1960	19°N 93°W		Disk UAP
Apr 1961	Bermuda Triangle		2UAPs 1 large, 1 small disk
1961	Remedios		3UAPs 1 large 2 small spheres
20th Nov 1964	18°N 66°W		Triangular UAP
1968-1969	Guantanamo Bay	Solid	Multiple UAPs emerged
Late Oct 1969	North of Cuba	Moderate	Massive UAP
17th Aug 1981	Las Villas		UAP cigar crashed
1984	Pinar del Rio		"Other" tall & Hairy
21st Dec 1993	Las Villas		Several "Others" winged men
1994	Guantanamo Bay		UAP Cigar
13th Dec 1995	Mayabeque		UAP spikey sphere

BRAZIL

Date	Place	Credibility	Details
26th Nov 1846	At Sea		8-10m UAP
9th Feb 1913	Cape San Roque		Skytrain of many UAPs
Mid Jul 1939	A spring in Minas Gerais		UAP 2 Tall "Others"
4th Dec 1949	Rio de Janeiro		Disk UAP 2 "Others" 1.65m abdu
28th Nov 1953	Rio Guapore – N Brazil	Solid	UAP 6 Human-like "Others"
9th Dec 1954	Linha Bela Vista		Disk UAP 2 "Others"
14th Dec 1954	Curitiba		3 UAPs
16th Jul 1956	Caraguatatuba		UAP disk out 2 tall "Others" fair
26th Jul 1956	Nr Rio de Janeiro, Brazil	Strong	2 UAPs
10th Sep 1956	Niteroi		Spherical UAP 2 "Others" 1.6m
15th Dec 1957	Floodplain by river		UAP DiskDome 7 "Others" 1.4m
Jan 1958	Lagoa Negra Lake		Disk UAP 2 Tall "Others" 3 1.4m
10th Jan 1958	Atlantic Ocean near Brazil	Moderate	Large UAP in/out
16th Jan 1958	Trinidade Island		5x 1 Disk UAP
1962	Canoas		"Other" heals crash victim 1.7m
31st Oct 1963	Peropava River, Brazil	Strong	UAP disk entered
26th Jul 1965	Carazinho		2 Ovoid UAP 5 1.5m "Others"
Oct 1965	Canhotiho		Tube UAP 2 "Others" 0.9m
26th Feb 1966	Agua Branca, Quipapa		Disk UAP 1 Tall, several 1.7m
16th Mar 1966	Nr Rio de Janeiro, Brazil	Strong	Oval UAP parachutes
1967	Dantana do Livramento		UAP carries VW Beetle 2km
Jun 1967	Sarandi		Sphere UAP 3 "Others" .8m Abdu
30th Jul 1967	Santa Catarina, Brazil	Certain	Cigar UAP dived
6th Feb 1969	Pirassununga		UAP disk "Other" 3x1.5m
12th Feb 1969	Pirassununga		2 "Others" 1.3m
27th Jun 1970	Rio de Janeiro,		UAP Disk landed on beach
30th Aug 1970	Rio de Janeiro, Funil Dam		UAP Guard shot at it, and was hurt
4th Apr 1971-74 C	Itaperuna		3 x Sphere UAP 3x1.0m "Others"
2nd Oct 1971	Rio de Janeiro,		UAP Disk
4th Jan 1975	Sao Luis		UAP Disk 11x1.3m Abductions
July 1977	Maranhao & Para		UAP waves lights & burns football
11th Mar 1978	Rio de Janeiro		UAP disk wave 12
16th Mar 1978	Pelotas		UAP Disk 3 tall fair "Others"
24th Nar 1978	Penalva		UAP Domed disk 3 "Others" 1m
12th May 1978	Pelotas		Dimed Disk UAP "Others" 1.1m
Jan 1979	Santa Cruz		Egg UAP
27th Jan 1979	Santa Cruz		UAP attempted abduction
Nov 1979	Santo Antonio		UAP attempted abduction
1981	Santa Catarina		UAP Reptile "Other" abduction
6th Mar 1982	Morenaho		Cylinder UAP over football game
Oct 1982	Antonina		UAP landed 3 "Others" 1.9m
12th Aug 1983	Maraponga lake		UAP Disk abduction, dreams
2nd Jul 1985	Curitibar		Abduction to Rio de Janeiro
Nov 1988	Carnaubinha		Sphere UAP 2 "Others" 1.5m

BRAZIL (Continued)

Date	Place	Credibility	Details
1991	Acari		UAP attempted abduction
May 1991	Campo Redondo		House sized UAP near abduction
5th Mar 1992	Ceara interior		Bus sized UAP 5 "Others" 1.3m
17th Apr 1996	Belem		2 UAPs 8 "Others" 2.0m+
19th May 1986	Rio de Janeiro		Multiple UAPs seen & Radar
21st Aug 1996	Brazilia dam		UAP
5th Oct 1996	Sao Jose do Norte		Giant pyramid UAP
15th Jun 1997	Lago Oeste		Daytime UAP 30cm diameter
20th Sep 2010	Ceara		Circular UAP, many lights
25th Oct 2010	Belem		UAP
14th Nov 2010	Ceara		UAP

ARGENTINA

Date	Place	Credibility	Details
1870	Buenos Aires		Lizard Man found in cavern
1897	White Lake Patagonia		Long-necked lake monster
1905	River inSantiago		Blond forecast floods
24th Dec 1930	Bahia Blanka		Light over several entities
1943	Lincoln, Buenos Aires		2 "Others" 1.2 m
1953	Pehuajo		UAP disk 1.5m "Other"
20th Sep 1954	Brasen		UAP disk 1,4 m "Other"
4th Dec 1954	Coroonel Pringles		UAP disk, Dwarf
Aug 1956	Balneario Norte		UAP disk Tall "Other" Abduc
1957	Las Juanitas		"Others" enter UAP Disk
Feb 1959	Bahia Blanca		Abductee taken 1,155 km
25th Dec 1959	Ezeiza		80 cm "Other"
1960	Villa Crespo		Lizard Men "Others" tunnel
1961	Mar del Plata		15 "Others" Abducting
Sep 1961	Rio Salado		UAP disk 2 1.5m "Others"
1961	Mar del Plata		UAP Disk 3 "Others"
May 1962	Pehuajo		UAP 3 short "Others"
1962	General La Madrid		UAP bald "Other"
1962	Mar del Plata		2 "Others" at baby's bedside
Feb 1963	Ranelach		"Other" with fangs attacking
Nov 1964	Lanus		Short "Other" big head no ears
20th Jul 1965	Quinas		UAP egg. 2 Nordics
20th Aug 1965	Mar del Plata		UAP 2 orange "Others"
22nd Oct 1965	Rancaua		2 short bighead "Others"
1967	Olavarria		2 Tall "Others"
Mar 1968	Altura Maipu		Abduction 6400 km to Mexico
Apr 1968	Mar del Plata		1 Tall "Other"
22nd May 1968	La Florida		"Other" 1m green, big ears
4th Jun 1968	Buenos Aires		1 Nordic "Other"
14th Jun 1968	Mar del Plata		UAP Triangle 2m blond "Other"
20th Jun 1968	Avellaneda		14 giant "Others"
3rd Jul 1968	Ollavaria		1 2m blond & 1 short "Other"
9th Jul 1968	La Plata		1 Tall "Other"
26th Jul 1968	Mar del Plata		3 "Others" Lights on helmets
26th Jul 1968	Ollavaria		UAP oval. 2 tall "Others"
30th Jul 1968	Mar del Plata		2 "Others" Lights on helmets
6th Aug 1968	Canuelas		UAP disk 1 Tall "Other"
31st Aug 1968	Bahia Blanca		1 Tall "Other"
8th Jan 1972	Pergamino		2 short "Others" big heads
16th Mar 1972	Azul		2 "Others" 1.2m & 1.7m
27th Aug 1972	Bahia Blanca		UAP "Other" with big head
4th Oct 1972	Burzaco		5-10 Tall "Others"
28th Nov 1972	Tres Arroyes		1 Tall "Other"
30th Dec 1972	Tres Arroyes		UAP 1 medium "Other"

ARGENTINA (Continued)

Date	Place	Credibility	Details
Jan 1973	Buenos Aires		1 Tall "Other"
Jun 1973	Villa Martelli		1.7m reptilian "Other"
Summer 1973	El Cholo		UAP 3 0.7m "Others"
29th Oct 1973	General Pinto		UAP 3 Nordic "Others"
Fall 1973	Carmen De Arco		2x1.5m "Others"
15th Jan 1974	Mar del Plata		5 "Others" in masks exit sea
25th Jun 1974	Tandil		Tall "Other" with small head
9th Jul 1974	Tres Arroyas		Tall "Other" Lizard
5th Jan 1975	Bahia Blanca		Tall "Other" small headed
23rd Nov 1977	Capital Federal		2 "Others" 1.4m green
Jan 1978	Buenos Aires		Tall "Other"
28th Jun 1978	Sarandi		"Other" 1.4m pointed ears
31st Aug 1978	Necochea		2 "Others" 0.7m
Oct 1978	Road in Buenos Aires		UAP 4 Tall "Others"
28th Jan 1980	San Carlos de Bariloke		UAP Disk 3 Tall "Others"
Jan 1980	Mar del Plata		UAP 3 "Others" 1.7m lights
5th Jun 1980	La Plata		UAP cigar 7 "Others" lights
2nd Aug 1980	Mar del Plata		UAP floating 1.7m "Other"
1982	Lincoln		1.2m "Other"
1982	Veronica		UAP 1.8m "Other"
Summer 1982	San Miguel del Monte		2.5m Tall "Other"
25th Dec 1982	La Plata		2 small "Others"
1983	Capital Federal		3 small "Others" Big Eyes,
Jan 1984	Ezeiza		2 tall "Others" flying
Jan 1984	BA Proven		Vaporous "Other" Tall green
3rd Apr 1984	Barrio Coghla		10 "Others" 0.6m UAP sphere
12th Dec 1984	Pilar		4 "Others" 0.5m flying
Feb 1985	Azul		2 "Others" 0.5m
7th Dec 1985	Buenos Aires		Tall "Other"
1985	Buenos Aires		"Others" 6' felines x 100
Feb 1986	Don Torcinato		"Other" Giant bat with horns
Summer 1986	Benavidez		Tall Blond "Other"
Jan 1988	Mar del Plata		UAP Oval "Other" 0.5m
4th Feb 1988	Necocha		2 "Others" Tall blond
24th Aug 1988	Mar del Plata		2 "Others" 0.6m
14th Oct 1988	Pergamino		5-7 "Others" 0.7m
6th Nov 1988	Pergamino		6-7 "Others" Grey Big head eyes
1988	Mar del Plata		"Other" 1m
Feb 1990	Punta Indio		2 "Others" 0.9m gnomes
Mar 1990	Villa Devota		"Other" 0.8m big head, eyes
31st Aug 1990	Berisso		2 "Others" 1.0m
Sep 1990	La Plata		2 "Others" 1.0m
Feb 1991	Lake in Patagonia		"Other" Tall
Nov 1991	Buenos Aires		"Other" Tall blond
Jun 1992	Villa Devota		"Other" 0.8m large head
Feb/ May1992	Villa Devota		"Other" big head, eyes

ARGENTINA (Continued)

Date	Place	Credibility	Details
Dec1992	Colonia Urquiza		2 small "Others"
Jan 1994	Lanus		Small "Other"
24th Apr 1994	Olavarria		19 "Others" Tall
1st Jan 1994	Bandero		"Other" Yeti-like
11 Feb 1995	San Antonio Ouest		2 "Others" 1 short, 1 heavy-set
15th Nov 1995	Mar Del Plata		3 "Others" Large
Feb 1996	Carlos Spigazzini		"Other" Canine biped with claws
May 1996	Rio Parana		"Other" Lizard man
19th Mar 1999	Moreno		"Other" Short big head & eyes
Jul 1999	Merdes		"Other" Orange skinned man
25th Aug 1999	Trenque Laquen		5 "Others" Tall 2.5m
Nov 1999	Trenque Laquen		"Other" Tall

SOUTH AFRICA

Date	Place	Credibility	Details
18th Jan 1914	Transvaal		UAP Cigar searchlight
19th Aug 1914	Orange Free State		UAP Cigar searchlight
21st Aug 1914	Mafeking		UAP Cigar searchlight
Sept 1914	Natal		UAP
Oct 1928	Transvaal		2 UAPs spheres
Summer 1946	Johannesburg		UAP Disk 2 "Others" 2.1m
21st Feb 1950	Durban		UAP cigar
27th Dec 1954	Natal		UAP Disk windows 20m
25th Jun 1960	Eastern Cape		UAP 9m long
14th Dec 1963	Transvaal		UAP 15m disk
15th Sep 1965	Transvaal		UAP Top 10m
2nd Jul 1972	Eastern Cape		UAP triangle very bright
5th Jul 1972	Eastern Cape		UAP red disk
16th Jul 1972	Orange Free State		UAP 1m disk
22nd Jul 1972	Natal		UAP 10m sphere
19th Aug 1972	Natal		"Other" tall hovering
20th Aug 1972	Eastern Cape		UAP Big silver sphere
21st Aug 1972	Natal		"Other" 3m
28th Aug 1972	Eastern Cape		UAP multiple lights
12th Nov 1972	Eastern Cape		UAP sphere
17th Nov 1972	Eastern Cape		2 "Others" 1.0m tall red
19th Nov 1972	Eastern Cape		2 UAPs domed
16th Jan 1973	Eastern Cape		UAP 30m
28th Jun 1973	Transvaal		UAP domed
31st Jul 1975	Eastern Cape		UAP like caravan
5th Mar 1976	Eastern Cape		UAP bright disk
2nd Oct 1978	Cape Province		Oval UAP 20m "Other" grey
24th Nov 1978	Cape Town		UAP large egg
3rd Jan 1979	Gauteng		UAP 6m egg "Other" 1.5m dark hair
Summer 1986	Durban		20 "Others" 1m
Apr 1993	Western Cape		UAP oval
30th Mar 1995	Transvaal		UAP disk hole in it
14th Jan 1997	Orange Free State		UAP grey sphere
6th Jun 1998	Durban		UAP oval
1998	Durban		Tall bald "Other"

PORTUGAL

Date	Place	Credibility	Details
29th May 1946	Lisbon		UAP sighted
24th Sep 1954	Castelo Branco		2 "Others" 2.2m giants
15th Oct 1954	Alentjo		2 "Others"
28th Aug 1957	Centro		UAP disk with dome
2nd Nov 1959	Alentejo		2 UAPs angel hair
22nd Apr 1960	Alentejo		UAP cigar
1962	Algarve		Tunnels found
1st Jul 1965	Azores airport		UAP stopped clocks
Aug 1967	Minho		UAP ellipse
Sep 1973	Douro		UAP lens 2m 2"Others" 1.7m antennas on heads
1974	Algarve		UAP disk "Other" very tall
3rd Jan 1977	Beira Alta		"Other" like tall box on legs
14th Jan 1977	Amarante		"Other" egg shaped head
Sep 1977	Algarve		4 UAPs
6th Jan 1978	Estremadura		"Other" 2.3m
Aug 1978	Algarve		Several "Others" Tall
Sep 1978	Dao-Lafoes		UAP light
15th Oct 1978	Algarve		UAP round 2 tall "Others"
25th Dec 1980	Oeiras		UAP oval
2nd Nov 1982	Estremadura		UAP disk
25th Sep 1983	Douro Litoral		3 UAPs
1986	Almada		"Other" 2.0m
4th Sep 1989	Lisbon		Abduction

AZORES

Date	Place	Credibility	Details
10th July 1948	Azores		UAP sighted
26th Sep 1952	400 miles NNW of Azores		UAP 2 lights near plane
20th Sep 1954	Azores airport		UAP 3m, "Other" 1.8m
16th Aug 1956	Azores airport		UAP light near plane
Aug 1967	Azores		2 "Others" 2.2m
2nd Feb 1968	Azores		UAP Oval 4"Others"
15th Oct 1976	Azores		UAP circular

CANARY ISLES

Date	Place	Credibility	Details
1912	Tenerife		Passageway found
Sep 1969	Gran Canarias		UAP huge clear sphere
Summer 1973	Gran Canarias		UAP Box shaped
Jan 1975	Las Palmas		5 UAPs lights
Aug 1975	Gran Canarias		"Other" 1.4m big head & ears
21st Dec 1975	Tenerife		UAP circular
22nd Jun 1976	Gran Canarias		UAP Sphere 15m
Feb 1977	Tenerife		UAP 100m sphere emerged 4"Others" tall blond
11th Apr 1978	Gran Canarias		UAP cigar
Sep 1978	La Graciosa		"Other" very tall
3rd Mar 1979	Tenerife		UAP pear shaped
Spring 1980	Tenerife		"Others" tall thin
22nd Apr 1989	Tenerife		UAP Huge sphere entered sea
7th Feb 1990	Lanzarote		UAP huge red sphere landed
Dec 1990	Tenerife		UAP Disk
12th Oct 1992	Tenerife		UAP crash
Nov 1992	Tenerife		"Other" 1m furry cat eyes
1st Jul 1997	Tenerife		"Others" winged
7th Jul 1997	Tenerife		UAP 30m saucer "Others" 1.2m
Jan 1999	Tenerife		UAP domed disk, Tall, long-haired "Others"

NORWAY

Date	Place	Credibility	Details
Summer 1869	Sor-Trondelag		UAP sighted
13th Aug 1897	Melo		2 UAPs
1st Apr 1908	Notodden		UAP crashed Egg shaped
Summer 1915	Sulitjelma		UAP landed, bell. 2 “Others”
Jan 1934	Multiple sites		UAPs – Ghost Fliers
31st Mar 1934	Sandnessjoen		Huge UAP
28th Jan 1937	Oftofjorden		Submersible UAP
14th Mar 1942	Banak		UAP cigar, radar & visual
1946	Scandinavian Lakes	Certain	UAPs diving into lakes
Aug 1947	Lake Djupsjoen	Solid	UAP blue egg entered
1950	Malselv		UAP disk “Other” contact
20th Sep 1952	Operation Mainbrace		UAP sphere photo
Oct 1952	Lagen River	Moderate	UAP entered, saucer
28th Nov 1953	Nr Oslo		UAP disk follows car
13th May 1954	Norbotten		UAP sphere
20th Aug 1954	Mosjoen		UAP disk, Tall “Other”
Jul 1957	Halingdal		UAP in flight
4th Jan 1958	Stavanger		UAP landed, Tall “Other”
1st Jun 1958	Alta Fjord	Strong	UAP entered. Delta wings
23rd Dec 1959	Skomsvol		UAP cigar dropped 2 UAPs
24th Sep 1961	Baltic, Nr Poland	Solid	UAP emerged
Winter 1962	Lake Stordalsvatnet		UAP winged
28th Feb 1963	Atlantic, Nr Norway	Solid	UAP buzzed British ships
21st Aug 1963	Skjervoy		UAP oval. Taking in water
21st Aug 1965	Kvaenangen Fjord	Strong	UAP fishing wartime wreck
14th Dec 1966	Porsan Fjord		UAP entered
30th Mar 1969	Oslo		UAP follows car
6th Jun 1969	Husnes		UAP disk
Jan 1970	Tronstad		UAP disk
3rd Nov 1970	Southern Norway		UAP light blinds car driver
1st Jan 1972	Bergen Airport		7 UAPs seen by pilot
1st Jan 1972	Nedre Lerfoss Power Stn		UAP cigar shape
Summer 1972	Solvang		UAP disk
20th Nov 1972	Fjordane multiple		Submersible UAP
10th Aug 1974	Hamar		Submersible UAP
Nov 1975	Arendal		UAP dropped off 8 larva
7th May 1977	Undersaker		2 UAPs triangles
Aug 1977	Narvik		UAP disk, 2 “Others”
1981-2-3-4	Hessdalen		Multiple UAPs
Nov 1984	Skogveien		UAP contactee Arve
28th Oct 1985	Hengsle, Honefoss		Multiple small “Others”
9th Mar 1992	Hamar		UAP Small triangle
Oct 1992	Hardangervidda		UAP 2 short “Others”
12th Apr 1977	Hordaland		UAP small triangle
Oct 1998	Trondheim		Large UAP Tall “Other”

SWEDEN

Date	Place	Credibility	Details
Jun 1935	Dalarna		"Other" Little man
1941	Stockholm - King		3 "Others" 1.2m dwarfs
Spring 1943	Ostergotland		UAP disk
Aug 1943	Kristianstad		"Others" line of little ones
25th May 1946	Stockholm		Start Ghost Fliers barrage
11th Jul 1946	Vastagotland		UAP Sphere crashed
11th Jul 1946	Orebro		UAP 20m triangle
Aug 1946	Tarno		"Other"
28 Sep 1952	Dalarna		2 UAPs Cigars with windows
18th Aug 1949	Stockholm		UAP disk 15m
13th May 1954	Norrbotten		3 UAPs disk
8th Sep 1957	Vastagotland		2 "Others" Cyclops conical
9th Nov 1958	Uppland		UAP flattened sphere 17m
20th Dec 1958	Skania		UAP 5m disk "Others" slugs
29th Sep 1959	Smaland		UAP 4m oval "Others" teens
29th Oct 1965	Stockholm		5 "Others" small big heads
5th Mar 1967	Vastagotland		UAP cigar 25-40m long
23rd Aug 1967	Ostergotland		"Other" 1.3m small mouth
1967	Vojman River		UAP30m disk
15th Feb 1971	Norrbotten		"Other" 1.0m cap
30th Apr 1971	Jonkoping		UAP Disk 10m
3rd May 1973	Ostergotland		UAP Sphere
21st May 1973	Kristianstad		"Other" large black bird
22nd Jan 1974	Stockholm		Several "Others" 2m thin
14th Jan 1975	Ostergotland		UAP short wings
Aug 1975	Scockholm		2 "Others" short gray
7th Jan 1976	Jamtland		UAP pyramid
3rd Mar 1976	Nottbotten		7 UAPs disk
Aug 1976	Dalarna		"Other" dwarf v large ears
14th Nov 1976	Kolsva		UAP tropical helmet 6m
31st Jul 1980	Lapland		UAP cigar entered lake
1st Nov 1980	Orebro		UAP oval 12m
Summer 1981	Smaland		"Other" 1.0m on film
20th Aug 1983	Dalarna		UAP globe
Summer 1985	Helsingland		"Other" Tall & hairy
Oct 1993	Vaserbotten		UAP disk
17th Mar 1997	Vastra Gotland		UAP ball
27th Jul 1999	Varnland		UAP 5m oblong entered lake

RUSSIAN LAKES & RIVERS

Date	Place	Credibility	Details
1663	Robozero Lake, Vologda		Large flaming sphere UAP
19th Century	Lake Lagoda		Regular Underground Noises
1884	Lake Baikal		Large UAP Sphere
1930s	Lake Issyk Kul		3 Giant "Other" Skeletons
1938	Lake West Yaroslavl		Several "Others" large heads
1939	Lake Shaitan Kirov		"Others" Green 1,2m
1941	Zelyeny Island River Don		UAP dogfight 1 crashed 20m
Summer 1943	Lake Shaitan Kirov		"Others" Tall
16th Jun 1948	Lake Baskunchak		UAP cigar damaged jet
1950s - 1960s	Chersky, Kolyma River		5-7 Disk UAPs
1950	Chersky, Kolyma River		Disk UAP 3 days. stayed 2 hrs
Sep 52	Chulym River Khakassia		"Others" Short large heads
1955	Enisey River Krasnoyarsk		"Others" little green men, playing with children
21st Apr 1961	Lake Onega		UAP Egg shaped crashed
27th Apr 1961	Lake Onega, Karelia	Strong	UAP in and out
15th Jun 1962	Kabarga River		UAP red sphere
1963	Lake Labynkir, Yakutsk		"Other" Lake monster
17th May 1964	Lake Baikal		UAP sphere approaches post.
Autumn 1965	Lake Baikal		Gigantic cigar UAP
17th May 1966	Nizhnyaya River		UAP 30m cone 3 Tall "Others"
June 1966	Lake Kolyvansky		UAP 8m droplet, 3 "Others"
Winter 1966	Lake Ladoga, Karelia		UAP disk 8 humans killed
2nd Aug 1972	Vyatka River, Kirov		"Other" 2m Tall pointed ears
1973	Lake Onega		Triangular UAP
Jul 1974	Esaulovka River Siberia		2 "Others" Tall Abduction
Jul 1975	Charvak Reservoir		UAP sphere exited
Aug 1975	Kuril Islands		Abduction 2 weeks
Feb 1997	Lake Lagoda		Triangular UAP
May 1977	Lake Kemsk, Vologda		UAP fiery sphere exploded
20th Sep 1977	Lake Onrga, Karelia		UAP Disk 100m
30th Dec 1980	Valaam I. Lake Ladoga		"Other" Tall
Summer 1982	Lake Baikal		Tall "Others" underwater
Aug 1988	Lake Lovozero Murmansk		"Other" Tall hairy
1982	Lake Sarez		UAPs enter and leave
13th Feb 1989	Lake Kopansky, Leningrad		UAP 30m disk windows
June 1989	Lake Zumerkiy, Vologda		2 Grey-green "Others"
Aug 1989	Lake Lagoda, Valaam I		Large Spherical UAP
19th Jan 1990	Lake Khnaka, Primorsky		UAP Disk landed on lake

RUSSIAN LAKES & RIVERS (Continued)

Date	Place	Credibility	Details
Feb 1990	Lake Issyk Kul		Large UAP sphere split into 4
21st Mar 1990	Amur River		UAPs Red Sphere, Cigar
30th Apr 1990	Lake Torbeevo		UAP 5m Sphere
1990s	Lake near Leningrad		Hang-glider - 3 Big "Others"
20th Jun 1990	Lake Issyk Kul		Triangular UAP above
17th Aug 1990	Lake Malyi Byadan		UAP 6m disk 2 Tall "Others"
13th Sep 1990	Lake Vselug Tver		UAP Sphere
1994-96	Lake Sukhodol'skoye		Spherical UAPs
21st Jul 1997	Caspian Sea		UAP entered sea
1990-2000	Lake Cheremenetskoye		Frequent UAPs
Spring 2009	Lake Baikal		2 orange UAP spheres

THE GREAT LAKES

Date	Place	Credibility	Details
27th Jan 1895	Lake Michigan	Weak	Large UAP emerged
Summer 1950	Flambeau Lake, Wisconsin	Weak	UAP emerged
26th Aug 1952	Seneca Lake, New York	Moderate	UAP entered
July 1955	Lake Ontario	Solid	UAP emerged
17th Sept 1955	Ticitus Reservoir, New York	Strong	UAP above surfa
20th July 1958	Small Lake, Michigan	Certain	UAP entered
14th Oct 1961	Lake Superior	Certain	Cigar UAP entered
7th Sep 1962	Montrose		Domed UAP
15th Sep 1962	Oradell Reservoir, New Jersey	Certain	UAP landed on water
August 1964	Upper Nemahbin L, Wisconsin	Solid	UAP in and out
Jan, Oct 1966	Wanaque Reservoir, New Jersey	Certain	Multiple UAPs above
June 1966	Straits of Mackinac, Michigan	Solid	2 UAPs emerged
20th June 1969	Southern Lake Michigan	Certain	UAP in and out
29th Jul 1989	Ludington		5 discs emerge
22nd May 2005	Michigan		Large irregular shape
1st Mar 2009	Oakville		Massive UAP
12th Sep 2009	Lake Huron		Entered Lake
15th Aug 2011	Bay City		Dull rectangle
17th Oct 2012	Coloma		6 Orbs emerge
14th Sep 2014	Lake Michigan		3 Orbs leave lake
24th Mar 2015	Toronto		V-shaped 5-7 lights
22nd Aug 2015	Hamilton		Multiple Lights
28th Nov 2016	Green Bay		Triangular UAP
9th Apr 2018	Benton Harbour		Cigar Shape
5th Sep 2018	Wasaga		Black Triangle

NORTHERN ITALY

Date	Place	Credibility	Details
1845	Tuscany		UAP huge disk
1864	Tuscany		UAP huge fiery globe
13th Dec 1884	Lombardy		UAP crash
Jun 1905	Emilia-Romagna		"Other" Flying Woman
Jul 1910	Tuscany		"Other" Giant flying bat
Dec 1919	Veneto		"Other" Luminous Spectre
1927	River Po, Veneto	Moderate	UAP in & out
Summer 1929	Emilia Romagna		UAP orange ball
Summer 1929	Tuscany		UAPs multi orange balls
13th Jun 1933	Lombardy		UAP crashed & studied
28th Nov 1942	Piedmont		UAP 65m long
13th Mar 1945	Tuscany		100 UAPs orange balls
1st Sep 1946	Marche		UAP into sea
14th Aug 1947	Friuli		3"Other" 1m green 8 fingers
3rd Apr 1948	San Martino, Lombardy		UAP & "Other"- 1.7m'
28th Oct 1049	Marche		20m UAP cigar
31st Mar 1959	Emilia Romagna		EAP Egg into sea
24th Apr 1950	Lombardy		UAP & 4 "Others" Helmet
1951	Piedmont		Huge UAP
31st Jul 1952	Musinè		UAP landed
18th Nov 1952	Venrto		UAP Disk with dome
19th Sep 1954	Rome		UAP Cigar
18th Oct 1954	Lombardy		UAP Short "Other"
19th Oct 1954	Tuscany		UAP 2 "Others"
27th Oct 1954	Tuscany Football Stad		Cigar UAP over thousands
28th Oct 1954	Rome		3 UAP flying fast
6th Nov 1954	Rome		Fleet UAPs
8th Nov 1954	Lombardy Stadium		UAP "Others" helmets
14th Nov 1954	Liguria		Cigar UAP 3 "Others"
15th Nov 1954	Rome		Fast UAPs
Sep 1955	Emilia Romagna		"Other" 1.0m
Apr 1956	Marche		2 "Others" 2.5m 1.0m
Jun 1958	Lombardy		UAP Box
15th Mar 1959	Turin Airport Piedmont		UAP Disk
3rd Jun 1961	Nr Savona, Adriatic Sea	Solid	UAP emerged
9th Apr 1962	Piedmont		UAP Disk
9th Nov 1962	Emilia Romagna		UAP landed 2 "Others"
17th Dec 1962	Milan		UAP landed 2 "Others"
4th Jan 1963	Rome		UAP disk with dome
20th Aug 1963	Rome		UAP domed disk
1st Nov 1964	Mount Musinè		UAP Landed
Jul 1967	Lombardy		UAP Disk 6 "Others"1.2M

NORTHERN ITALY (Continued)

Date	Place	Credibility	Details
15th Nov 1967	Tuscany		UAP delta shaped
22nd Aug 1968	Latina		UAP Disk with dome
24th Sep 1970	Piedmont		Triangular UAP
4th Jul 1973	Liguria		Round UAP ball of fire
30th Nov 1973	Turin Airport, Piedmont		Large UAP
5th Dec 1973	Piedmont		UAP radar and thousands
16th Apr 1974	Piedmont		UAP Ring with Cockpit
15th Feb 1975	Tuscany		2 “Others” Tall
Jun 1975	Emilia Romagna		“Other” Giant lizard
13th Jan 1976	Liguria		4 “Others” 1.6m
28th Mar 1976	Marche		Oval UAP left sea
14th May 1976	Liguria		“Other” 1.3m
10th Sep 1976	Liguria		“Other” 1.2m
1st Jul 1977	NATO Base Friuli		UAP bright light
6th Oct 1977	Tuscany		3 UAPs half moon size
23rd Apr 1978	Tuscany		“Other” 1.0m
24th Oct 1978	Marche		Large UAP up and down
18th Dec 1978	All Over Italy		Multiple UAPs
31st Dec 1978	Marche		UAP sphere up from sea
16th Jan 1979	Piedmont		10m UAP sphere
3rd Feb 1979	Liguria		“Other”1.2m helmet light
7th Feb 1979	Lombardy		“Other” 2.0m long arms
4th Oct 1979	Viggiu, Liguria		2 “Others” dwarfs <1m
14th Jun 1980	Liguria		2 “Others” 2.0m
8th Nov 1981	Veneto		UFO oval 2m “Other” no neck
Jul 1983	Tuscany		UFO 60m sphere
9th Oct 1984	Tuscany		Elliptical UAP. “Other”
7th Jun 1986	Piedmont		“Other” Large Red Hair
Summer 1986	Veneto		“Other” Tall Reptilian
15th Aug 1986	Veneto		UAP Domed disk
27th Sep 1986	Emilia Romagna		UAP Huge with wings
28th Oct 1986	Tuscany		UAP large triangle
16th Feb 1988	Piedmont		UAP disk “Other” 2.0m
20th Apr 1992	Veneto		UAP cigar 2”Others”
Jun 1993	Marche		“Other” Dwarf big feet
10th Jul 1993	Lombardy		5“Others” Luminous
3rd Feb 1994	Marche		“Other” 1.5m big feet
22 Jul 1996	Veneto		“Other” 1.0m
9th Nov 1996	Veneto		3 “Others” small grey big head
1st May 1998	Lombardy		“Other” big hairy
18th Sep 1998	Liguria		UAP blue oval
6th Jun 2010	Vatican City		Triangular UAP
27th Apr 2013	Province of Verona		UAP

INDEX

REFERENCES

[1] UFO Friend or Foe? by Martin Thomas Pub 2025
[2] UNESCO World Heritage Convention. The List
[3] Scientific & Esoteric Encyclopaedia of UFOs, Aliens & Extraterrestrial Gods V4 p156 by Maximillien de Lafayette pub NY 2014
[4] The Alien Agendas pp 11-21 by Richard Dolan Pub Richard Dolan Press, New York 2020
[5] Wikipedia
[6] Alaska's Deadly Triangle p32 by Betsey Lewis 2023
[7] Stargates P27 by Betsey Lewis 2021
[8] UFO Sightings Hotspot 2024
[9] Irish News UFO footage of Alaska 2024
[10] International Business Times 426515 2013
[11] Wikipedia
[12] Wikipedia
[13] Wikipedia
[14] Stargates P61 by Betsey Lewis 2021
[15] Encyclopaedia of Alien Races by Màesta Pastore
[16] The Naked Ape by Desmond Morris Pub Jonathan Cape London 1967
[17]Man and Dolphin: Adventures of a New Scientific Frontier Lilly, Dr John C. Lilly 1961 ·
[18] Nature Magazine 12 August 2015
[19] Consciousness of octopuses—on their own terms. Jenifer Mather *Animal Sentience 2025*
[20] Stargates P27 by Betsey Lewis 2021
[21] A History of USOs Vol1, Beginnings to 1969 by Richard Dolan Pub Richard Dolan Press Rochester USA 2025
[22] The Secret Space Program and Breakaway Civilisation P48 by Richard Dolan Pub Richard Dolan Press Rochester USA 2025
[23] Wikipedia
[24] Encyclopaedia of Alien Races by Màesta Pastore P151
[25] The Alien Agendas pp 14 by Richard Dolan Pub Richard Dolan Press, New York 2020

[26] UFOs, Humanoids and Strange Phenomena of Africa, Asia and the Middle East Geotge Mitrovic
[27] Russia's USO Secrets P111, Stonehill & Mantle Flying Disk Press 2020
[28] UFOs, Humanoids and Other Strange Phenomena of Russia. George Mitrovic
[29] The Alien Colonisation of Earth's Waters by Debbie Zieglmeyer Pub UnX Media Missouri 2021
[30] Russia's USO Secrets P111, Stonehill & Mantle Flying Disk Press 2020
[31] The Extraterrestrial Species Almanac p37 Craig Campobasso Red Wheel
[32] The Alien Colonisation of Earth's Waters by Debbie Zieglmeyer Pub uUnX Media Missouri 2021
[33] Amazing Encounters with UFOs of West Coast North America, George Mitrovic
[34] The Alien Colonisation of Earth's Waters P163 by Debbie Zieglmeyer Pub UnX Media Missouri 2021
[35] Amazing Encounters with UFOs of West Coast North America, George Mitrovic
[36] The Alien Colonisation of Earth's Waters P175 by Debbie Zieglmeyer Pub UnX Media Missouri 2021
[37] A History of USOs Vol1, Beginnings to 1969 by Richard Dolan Pub Richard Dolan Press Rochester USA 2025
[38] UFOs, Humanoids and Strange Phenomena of Central America, the Caribbean and Mexico as well as the Atlantic Ocean, George Mitrovic.
[39] UFOs, Humanoids and Strange Phenomena of Argentina, Chile, Peru and Uruguay. George Mitrovic
[40] UFOs in Central and South American Air Spa. John Scott Chace 2020
[41] UFOs, Humanoids and Strange Phenomena of Argentina, Chile, Peru and Uruguay. George Mitrovic
[42] UFOs in Central and South American Air Space. John Scott Chace 2020
[43] Amazing Encounters with Monsters and Other Mysteries of Australia, New Zealand, the Pacific and Antarctica by George Mitrovic
[44] Amazing Encounters with Monsters and Other Mysteries of Australia, New Zealand, the Pacific and Antarctica by George Mitrovic
[45] UFOs Down Under by Barry Watts, Pegasus Education Group 2017
[46] A History of USOs Vol1, Beginnings to 1969 by Richard Dolan Pub Richard Dolan Press Rochester USA 2025
[47] Wikipedia

---

[48] Wikipedia
[49] Blue Mountains Triangle. Rex & Heather Gilroy URU Publications 2006
[50] News.com.au April 27 2013
[51] Britannica
[52] News on Japan article 145271
[53] Russian USO Secrets by Stonehill & Mantle, Flying Disk Press 2020
[54] UFOs, Humanoids and Strange Phenomena in the Ukraine, Belarus, Moldavia and Scandanavia. George Mitrovic
[55] UFOs, Humanoids and Other Strange Phenomena of Russia. George Mitrovic
[56] Amazing Encounters with Monsters and Other Mysteries of Australia, New Zealand, the Pacific and Antarctica by George Mitrovic
[57] Amazing Encounters with UFOs of East Coast North America Vol 2, George Mitrovic
[58] Amazing Encounters with UFOs of Canada, George Mitrovic
[59] UFOs, Humanoids and Strange Phenomena of Central America, the Caribbean and Mexico as well as the Atlantic Ocean, George Mitrovic.
[60] Amazing Encounters with UFOs of East Coast North America Vol 1, George Mitrovic
[61] Amazing Encounters of the South. George Mitrovic
[62] UFOs, Humanoids and Strange Phenomena of Central America, the Caribbean and Mexico as well as the Atlantic Ocean, George Mitrovic.
[63] A History of USOs Vol1, Beginnings to 1969 by Richard Dolan Pub Richard Dolan Press Rochester USA 2025
[64] The Alien Colonisation of Earth's Waters by Debbie Zieglmeyer Pub UnX Media Missouri 2021
[65] The Alien Colonisation of Earth's Waters P151 by Debbie Zieglmeyer Pub UnX Media Missouri 2021
[66] UFO Contacts in Brazil, by Thiago Luiz Tictti. Flying Disk Press 2019
[67] Alien Base p 224 by Timothy Wood pub Century London 1988
[68] UFOs, Humanoids and Strange Phenomena of Argentina, Chile, Paraguay, Peru and Uruguay, George Mitrovic.
[69] Alien Base p 224 by Timothy Wood pub Century London 1988
[70] UFOs, Humanoids and Strange Phenomena of Africa, Asia and the Middle East Geotge Mitrovic
[71] UFOs, Humanoids and Strange Phenomena in the Ukraine, Belarus, Moldavia and Scandinavia. George Mitrovic

[72] UFOs, Humanoids and Other Strange Phenomena of Russiaa. George Mitrovic
[73] Russian USO Secrets by Stonehill & Mantle, Flying Disk Press 2020
[74] The Alien Colonisation of Earth's Waters P151 by Debbie Zieglmeyer Pub UnX Media Missouri 2021
[75] Wikipedia
[76] UFOs, Humanoids and Strange Phenomena in Italy & Eastern Europe. George Mitrovic
[77] UFO Contacts In Italy Volume One
[78] UFO Investigations in Italy, Carlo Pirola Picadilly, NY City 2019
[79] The Extraterrestrial Species Almanac Craig Campobasso Red Wheel
[80] The Illustrated Guide to Reported Alien Species P25 David McDaniel 2017
[81] The Extraterrestrial Species Almanac p157 & 167 Craig Campobasso Red Wheel
[82] The Extraterrestrial Species Almanac p57 Craig Campobasso Red Wheel
[83] The Illustrated Guide to Reported Alien Species P51 David McDaniel 2017
[84] The Extraterrestrial Species Almanac p29 Craig Campobasso Red Wheel
[85] The Extraterrestrial Species Almanac p37 Craig Campobasso Red Wheel
[86] The Extraterrestrial Species Almanac p25 Craig Campobasso Red Wheel
[87] The Illustrated Guide to Reported Alien Species P27 David McDaniel 2017
[88] The Extraterrestrial Species Almanac P187 Craig Campobasso Red Wheel
[89] UFO Friend or Foe? P38 by Martin Thomas Pub 2025
[90] The Alien Agendas P175. By Richard Dolan. Richard Dolan Press 2020
[91] UFOs, Humanoids and Strange Phenomena of Argentina, Chile, Paraguay, Peru and Uruguay, p10 George Mitrovic.
[92] Encyclopedia of Alien Races P164 Maesta Pastore
[93] UFOs and Aliens 2 p141 Carl Spartacus
[94] UFO Friend or Foe? by Martin Thomas Pub 2025
[95] UFOs and Aliens 2 p126 Carl Spartacus Dunstable
[96] Alien Underground Bases Gil Carlson Wicked Wolf Press 2015

www.ingramcontent.com/pod-product-compliance
Lightning Source LLC
LaVergne TN
LVHW050916210926
848482LV00012B/463